AF794991

Optoelectronic Tweezers for Manipulation at the Micro and Nano Scale

Optoelectronic Tweezers for Manipulation at the Micro and Nano Scale

Lin Feng, Menglu Tan, Ao Wang, and Jiaying Zhang

WILEY VCH

Authors

Lin Feng
Beihang University
XueYuan Road No.37
HaiDian District
Beijing
CH, 100191

Menglu Tan
Beihang University
XueYuan Road No.37
HaiDian District
Beijing
CH, 100191

Ao Wang
Beihang University
XueYuan Road No.37
HaiDian District
Beijing
CH, 100191

Jiaying Zhang
Peking University Third Hospital
North Garden Road No.49
HaiDian District
Beijing
CH, 100191

Cover Design: Wiley
Cover Image: © Chris Rogers/Getty Images

Typesetting: Straive, Chennai, India
Printing and Binding: CPI Group (UK) Ltd, Croydon, CR0 4YY

Library of Congress Card No.: applied for

British Library Cataloguing-in-Publication Data A catalogue record for this book is available from the British Library.

Bibliographic information published by the Deutsche Nationalbibliothek
The Deutsche Nationalbibliothek lists this publication in the Deutsche Nationalbibliografie; detailed bibliographic data are available on the Internet at <http://dnb.d-nb.de>.

Print ISBN: 978-3-527-35460-3
ePDF ISBN: 978-3-527-85008-2
ePub ISBN: 978-3-527-85007-5
oBook ISBN: 978-3-527-85009-9

© 2026 WILEY-VCH GmbH, Boschstraße 12, 69469 Weinheim, Germany

C9783527354603_241125

All rights reserved (including those of translation into other languages, text and data mining and training of artificial intelligence technologies or similar technologies). No part of this book may be reproduced in any form – by photoprinting, microfilm, or any other means – nor transmitted or translated into a machine language without written permission from the publishers. Registered names, trademarks, etc. used in this book, even when not specifically marked as such, are not to be considered unprotected by law.

The manufacturer's authorized representative according to the EU General Product Safety Regulation is WILEY-VCH GmbH, Boschstr. 12, 69469 Weinheim, Germany, e-mail: Product_Safety@wiley.com.

In view of ongoing research, equipment modifications, changes in governmental regulations, and the constant flow of information relating to the use of experimental reagents, equipment, and devices, the reader is urged to review and evaluate the information provided in the package insert or instructions for each chemical, piece of equipment, reagent, or device for, among other things, any changes in the instructions or indication of usage and for added warnings and precautions. While the publisher and the authors have used their best efforts in preparing this work, including a review of the content of the work, neither the publisher nor the authors make any representations or warranties with respect to the accuracy or completeness of the contents of this work and specifically disclaim all warranties, including without limitation any implied warranties of merchantability or fitness for a particular purpose. No warranty may be created or extended by sales representatives, written sales materials or promotional statements for this work. The fact that an organization, website, or product is referred to in this work as a citation and/or potential source of further information does not mean that the publisher and authors endorse the information or services the organization, website, or product may provide or recommendations it may make. This work is sold with the understanding that the publisher is not engaged in rendering professional services. The advice and strategies contained herein may not be suitable for your situation. You should consult with a specialist where appropriate. Further, readers should be aware that websites listed in this work may have changed or disappeared between when this work was written and when it is read. Neither the publisher nor authors shall be liable for any loss of profit or any other commercial damages, including but not limited to special, incidental, consequential, or other damages.

Contents

Preface

In this book, we embark on a journey to explore the fascinating world of Optoelectronic Tweezers (OET) and their applications in micro- and nanoscale manipulation. Starting from the evolution of nanotechnology, we delve into the development of optical micromanipulation technologies such as Optical Tweezers (OT) and the emergence of OET as a cutting-edge technology with superior performance in various aspects. The fundamental principles of OET, its significance, and its diverse applications across different fields are meticulously examined.

Comprising several chapters, this book covers the working mechanisms, operational principles, optical foundations, photovoltaic effects, and material selection processes in OET technology. Detailed insights into the components of OET devices, including standard OET and photovoltaic OET, are provided, along with a comprehensive analysis of manipulation forces and dielectrophoretic effects within OET chips. The intricate interplay of optical principles, light sources, and photovoltaic mechanisms in OET setups is elucidated, shedding light on the complex yet fascinating world of optical manipulation at the micro- and nanoscales.

Furthermore, this book delves into the practical applications of OET in biomedical fields, showcasing its efficacy in cell sorting, trapping, lysis, patterning, immunoreaction analysis, DNA transfection, and other crucial biological processes. The versatility of OET in manipulating a myriad of micro- and nano-objects, ranging from living organisms like cells and bacteria to non-living entities such as nanoparticles and microspheres, is explored in detail. Additionally, the integration of OET with complementary technologies like microfluidics, photocuring, electrowetting, and image recognition is highlighted through specific examples, demonstrating the synergistic potential of these combined approaches.

As OET technology continues to garner widespread interest in academia and industry, this book presents a collection of experiments and case studies that underscore its practical utility and innovative applications across diverse disciplines. By examining the current challenges and future directions of OET, we aim to provide valuable insights into enhancing optoelectronic material selection, refining device structures, and expanding the scope of applications through advanced techniques like 3D manipulation and automation. It is our belief that OET technology holds immense promise in revolutionizing fields such as materials science, biology, pharmaceuticals, and beyond.

We hope that this book serves as a comprehensive guide for readers seeking a deeper understanding of OET and its transformative potential in scientific research and technological advancements. Join us on this exciting exploration of OET technology, as we unravel its limitless possibilities and pave the way for a new era of precision manipulation at the micro- and nanoscales.

May this book inspire curiosity, spark innovation, and ignite a passion for discovery in the realm of OET. Let us embark on this enlightening journey together, as we unlock the boundless opportunities offered by OET technology!

1

Introduction

1.1 History Background and Brief Overview

In the latter part of the 20th century, nanotechnology experienced a remarkable evolution, transitioning from an abstract concept to a tangible science that revolutionized numerous scientific fields. This period was characterized by the advent of several groundbreaking nanotechnological tools and techniques, which allowed researchers to explore and manipulate matter at the atomic and molecular scales. During this period, nanotechnologies such as atomic force microscopy (AFM), magnetic tweezers (MTs), acoustic tweezers (ATs), optical tweezers (OTs), and others emerged successively, each with its own strengths.

The AFM, also known as the scanning force microscopy (SFM), represents a groundbreaking advancement in nanotechnology and materials science (Figure 1.1). As a high-resolution scanning probe microscope, AFM operates at the nanoscale, offering imaging capabilities that surpass the optical diffraction limit by an astonishing factor of 1000. This remarkable achievement has revolutionized our ability to observe and manipulate matter at the atomic level.

The development of AFM can be traced back to its precursor, the scanning tunneling microscopy (STM), which was conceived by Calvin Quate, Gerd Binnig, and Christoph Gerber at the IBM Zurich Research Laboratory in 1986. Building upon the principles of STM, AFM has evolved into one of the most indispensable tools in nanotechnology, serving critical functions in material manipulation, imaging, and precise measurement at the nanoscale.

The operational principle of AFM centers around a sophisticated detection system consisting of a microcantilever and an extremely sharp probe attached to its end. This configuration allows for the precise "sensing" of surface characteristics through the interaction between the probe and the sample surface. The system's precision is further enhanced by piezoelectric elements, which enable extremely accurate and controlled movements of either the sample or the scanner, typically in the range of nanometers.

One of the most significant advancements in AFM technology is the development of conductive cantilevers and specialized conductive AFM attachments. These innovations allow researchers to measure the current bias of samples, opening up new possibilities in electrical characterization at the nanoscale. More sophisticated AFM

Optoelectronic Tweezers for Manipulation at the Micro and Nano Scale, First Edition.
Lin Feng, Menglu Tan, Ao Wang, and Jiaying Zhang.
© 2026 WILEY-VCH GmbH. All rights reserved, including rights for text and data mining and training of artificial intelligence technologies or similar technologies. Published 2026 by WILEY-VCH GmbH.

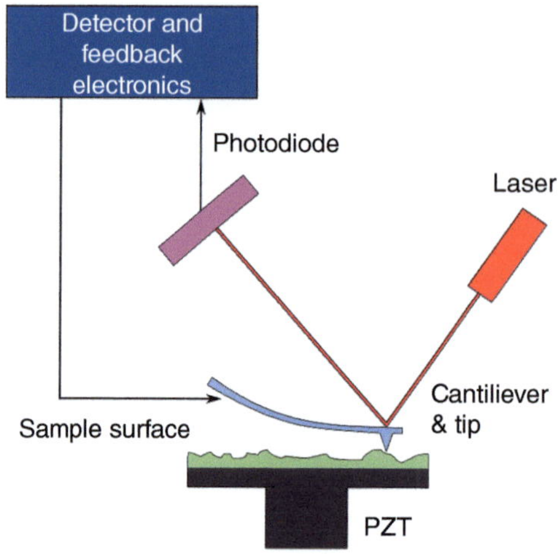

Figure 1.1 An AFM generates images by scanning a small cantilever over the surface of a sample. *Source*: OverlordQ's / Wikimedia Commons / CC BY-SA 4.0 / https://en.wikipedia.org/wiki/Atomic_force_microscopy#/media/File:Atomic_force_microscope_block_diagram.svg / last accessed by October 03, 2025.

instruments have pushed these capabilities even further, enabling the measurement of current on the probe itself. This capability allows for the assessment of a sample's conductivity and the observation of electron movement on the sample's underside. However, it is important to note that these advanced measurements present considerable technical challenges, and only a select number of research laboratories worldwide have successfully reported consistent and reliable data from such experiments.

The true power of AFM lies in its ability to sense and amplify the interaction forces between the sharp probe on the cantilever and the atoms of the sample being examined. This capability is what enables AFM to achieve its extraordinary atomic-level resolution. By precisely measuring these atomic-scale interactions, AFM provides researchers with unprecedented insights into the structure and properties of materials at the most fundamental level.

The applications of AFM span across numerous scientific disciplines, including materials science, nanotechnology, biology, and surface chemistry. Its ability to operate in various environments (vacuum, liquid, or air) and modes (contact, noncontact, or tapping) makes it an exceptionally versatile tool for nanoscale research and development. As the AFM technology continues to evolve, it promises to unlock even more possibilities in our understanding and manipulation of matter at the atomic scale, driving innovation in fields ranging from semiconductor technology to biomedical research [1].

Magnetic tweezers (MT) have become a significant and adaptable tool in biophysics and soft matter studies, facilitating the precise control and analysis of single biomolecules or polymers (Figure 1.2). These advanced instruments are engineered to apply specific forces and torques at the molecular level, offering scientists an unparalleled understanding of the mechanical attributes and behaviors of biological macromolecules.

The principal function of MTs is their capacity to assess foundational mechanical properties, such as the tensile strength of molecules or the forces involved in molecular interactions. This ability has rendered them, especially useful in single-molecule experiments, where they are predominantly used to examine the

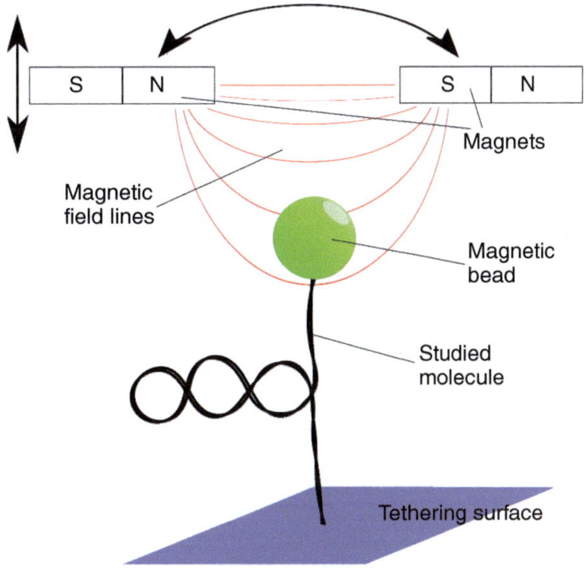

Figure 1.2 Typical configuration for magnetic tweezers. *Source*: Daniel Förster / Wikimedia Commons / CC BY-SA 4.0 / https://en.wikipedia.org/wiki/Magnetic_tweezers#/media/File:Magnetic_tweezers.svg / last accessed by October 03, 2025.

mechanical characteristics of biological macromolecules like DNA and proteins. In addition to these uses, MTs have proved beneficial across a variety of research domains, including the study of the rheology of soft matter systems and the exploration of force-dependent processes in living cells. The usual range of forces for these experiments extends from piconewtons to nanonewtons, making them perfectly suited for examining the delicate mechanical interactions that underpin biological functions.

One of the key advantages of MTs is their relatively simple and robust architecture, which has contributed to their widespread adoption as a biophysical tool. The experimental setup typically involves tethering the molecule of interest to a magnetic microparticle, which serves as the handle for manipulation. The MT system is equipped with precisely controlled magnets that generate the necessary magnetic fields to manipulate these particles in three-dimensional space. The position and movement of the magnetic particles are then monitored with high precision using advanced video microscopy techniques, allowing for the real-time tracking and analysis of molecular behavior.

The versatility of MTs extends their ability to perform both force spectroscopy and torque measurements, making them particularly valuable for studying the mechanical properties of DNA and proteins. For instance, they have been instrumental in elucidating the mechanisms of DNA supercoiling, protein folding, and the behavior of molecular motors. In soft matter research, MTs have been employed to investigate the viscoelastic properties of complex fluids and the mechanical response of biological membranes.

Recent advancements in the MT technology have further enhanced their capabilities, including improvements in force resolution, the development of multiplexed systems for parallel measurements, and the integration with other single-molecule techniques. These developments have opened up new possibilities for studying complex biological processes and material properties at the molecular level [2].

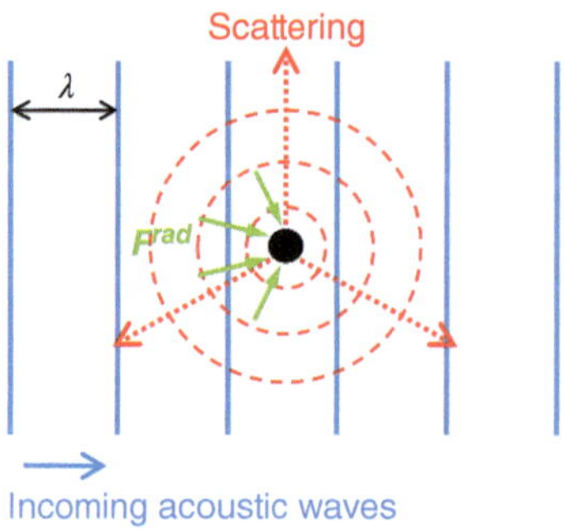

Figure 1.3 Acoustic radiation force on a small particle. *Source*: Rsla1 / Wikimedia Commons / CC BY-SA 4.0 / https://en.wikipedia.org/wiki/Acoustic_tweezers#/media/File:Figure1_zhangming.png / last accessed by October 03, 2025.

Acoustic tweezers (AT) represent a cutting-edge technology that harnesses the power of sound waves to precisely manipulate the position and movement of microscopic and sub-millimeter objects (Figure 1.3). This innovative approach to particle manipulation has opened up new possibilities in various scientific fields, particularly in biomedical research and nanotechnology. At its core, ATs carefully utilize controlled sound waves to create acoustic pressure fields that can influence the position of small objects. While the term "AT" strictly refers to a single-beam configuration, the broader concept encompasses two primary configurations: single-beam and standing-wave setups. The technology operates by manipulating the position of acoustic pressure nodes, which serve as stable points that attract and hold objects within a standing acoustic field. For effective manipulation, the target object must be significantly smaller than the wavelength of the sound used, typically ranging from microscopic particles to individual cells.

One of the most significant advantages of ATs is their noninvasive nature and safety for biological applications. Unlike other manipulation techniques that might involve potentially harmful forces or radiation, acoustic waves have been proven to be gentle and safe for biological objects. This characteristic makes ATs particularly valuable in biomedical research, where maintaining cell viability and integrity is crucial. Recent advancements in the AT technology have expanded its applications across various domains, including flow cytometry, where they have been integrated into systems to enable more precise cell sorting and analysis without the need for physical contact or labels. They have also proven effective in separating different cell types based on their physical properties, offering a label-free alternative to traditional separation methods.

Additionally, ATs can create stable traps for individual cells or groups of cells, allowing for long-term observation and manipulation. Researchers can now manipulate individual cells with unprecedented precision, enabling the study of cell mechanics, response to stimuli, and intercellular interactions. Beyond biological applications, ATs have shown promise in the precise arrangement and manipulation of nanomaterials, opening up possibilities in nanotechnology and materials science. The versatility of ATs stems from their ability to create complex acoustic fields that can be dynamically controlled. This capability allows for the manipulation of multiple objects simultaneously, as well as the creation of intricate patterns or arrangements of particles.

Moreover, the contactless nature of acoustic manipulation minimizes the risk of contamination or damage to delicate samples, making it an attractive option for various applications in cleanroom environments or sterile laboratory settings. As research in this field continues to advance, ATs are expected to play an increasingly important role in areas such as tissue engineering, drug delivery, and the development of advanced materials. The ability of the technology to manipulate objects at different scales, from individual cells to nanomaterials, positions it as a powerful tool in the rapidly evolving fields of biotechnology and nanotechnology [3].

Among them, OTs represent one of the techniques in optical micromanipulation. This field utilizes light to trap and control tiny objects (Figure 1.4). The principles of optical manipulation are based on interactions between light and matter, resulting in the creation of various forces for manipulation, such as optical gradient force, optical scattering force, optical torque, optical pulling force, and optical binding force. Research in optical manipulation technology has significantly advanced our understanding of the microscopic world across various scientific disciplines over the past five decades. The concept of optical micro-manipulation, later known as OTs, was introduced by Arthur Ashkin in the 1980s [4]. OT devices use focused laser beams to exert forces (typically in the range of piconewtons) on microscopic transparent objects. The area where the object is held is referred to as an optical trap, and the technique is known as optical trapping. This technology is utilized for moving cells or virus particles, shaping cells, and cooling atoms. Ashkin first explored the use of light radiation pressure to manipulate physical objects in the late 1960s and began working on the optical manipulation of microparticles [5]. In 1986, Ashkin pioneered the use of a single laser beam for trapping and manipulating

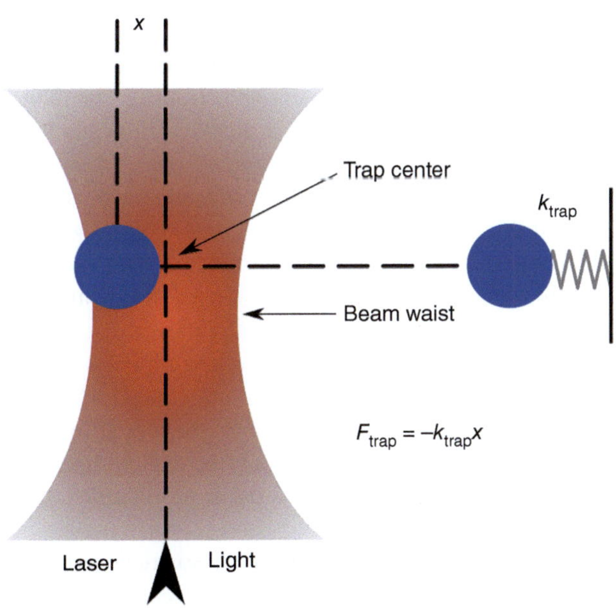

Figure 1.4 Dielectric objects are attracted to the center of the beam. *Source*: Cr4sZz / Wikimedia Commons / CC BY-SA 4.0 / https://en.wikipedia.org/wiki/Optical_tweezers#/media/File:Optical_trap_principle_formula_edit.svg / last accessed by October 03, 2025.

microparticles, leading to the invention of OTs [4]. OTs have been widely applied in physical, chemical, and biological sciences, revolutionizing research in the microscopic realm. In 2018, Arthur Ashkin was awarded the Nobel Prize in Physics for his work on OTs, highlighting the importance and impact of this technology [6]. Despite their utility, OT faces challenges. They are limited to manipulating small, transparent objects (ranging from nanometers to a few tens of microns) due to the small forces they exert. Additionally, OTs require high numerical aperture lenses and complex beam-shaping optics, and they can cause photothermal damage to biological samples. These challenges have restricted the widespread use of OTs to niche and high-end applications in research settings. Motivated by OTs and the need to address these challenges, significant research efforts have been directed toward developing alternative optical micromanipulation technologies.

A novel optical manipulation technique called opto-thermoelectric tweezers (OTETs) has been developed, harnessing thermoelectric forces directed by light. OTETs operate at significantly lower power levels compared to the traditional OTs, as they rely on temperature gradients rather than absolute temperature changes. The adaptability of OTETs is significantly boosted by the universal thermophoretic effect in solutions, which allows for the handling of a diverse range of materials, including polymers, metals, semiconductors, and dielectric nanostructures of varying dimensions and configurations. In addition, OTETs provide the ability to maneuver metal nanoparticles over a wide range of adjustable operational wavelengths, delivering versatile platforms for real-time optical measurements while minimizing the overlap between manipulation and measurement beams. By designing optothermal substrates with nanoscale heat sources, OTETs achieve spatial resolution that surpasses the diffraction limit in nanoparticle manipulation [7].

The theoretical idea of merging optical trapping with plasmonics was first suggested in 1997. As outlined in scholarly journals, a substantial radiation force can be exerted on a dielectric nanoparticle using the intense electromagnetic (EM) field present near the surface of a metal tip. The theoretical framework posits that nanoparticles can be effectively trapped optically when situated close to such a surface. Following this theoretical insight, numerous research teams have carried out experimental studies on plasmonic optical trapping. This method offers several benefits over traditional optical trapping. One primary advantage is that the EM enhancement effect of plasmons reduces the laser light intensity necessary for stable optical trapping compared to standard OT. Moreover, the movement of the trapped particle is constrained within the plasmonic nanospace, which is much smaller than the diffraction limit of light. Additionally, plasmonic optical trapping holds promise for integration with conventional plasmonic applications, including enhancing photochemical reactions and creating highly sensitive chemical and biological sensors [8]. These integrations could pave the way for innovative plasmonic applications.

Xie et al. developed an innovative approach termed "optoacoustic tweezers" for the dynamic aggregation and arrangement of particles and cells within a microfluidic environment. This method, which is both programmable and suitable for biological systems, leverages opto-thermally induced and acoustically activated surface bubbles. Optoacoustic tweezers allow for the clustering of particles or cells at designated spots inside a microfluidic chamber without the necessity of permanent structures, making it an excellent option for crafting adaptable and complex cellular patterns. The technique has demonstrated superb biocompatibility and can be seamlessly integrated with other microfluidic components [9]. Fiber optic tweezers (FOTs) have emerged as a particularly valuable tool due to their straightforward setup and seamless integration with lab-on-a-chip systems [10]. The inherent flexibility of optical fibers allows for the precise delivery of laser light to targeted areas, making FOTs an ideal choice for applications requiring compact and portable manipulation systems. Building upon the foundation of the traditional OT, scanning optical tweezers (SOT) were developed by incorporating advanced laser scanning technologies [11]. This innovation enables the dynamic trapping and precise movement of microparticles through the use of scanning mirrors or acousto-optical deflectors (AODs) [12]. The AOD-based systems, in particular, offer rapid beam steering capabilities, allowing for the real-time manipulation of multiple particles with high spatial resolution. This technology has proven particularly useful in biological applications where the dynamic control of cellular components is required. The field witnessed another significant leap with the development of holographic optical tweezers (HOTs), which employ spatial light modulators to create complex interference patterns [13]. By applying the holographic technology, HOT can split a single laser beam into multiple independent beams, enabling the simultaneous manipulation of numerous targets in three-dimensional space. This capability has opened new possibilities in colloidal physics and biological research, where multiple interactions need to be controlled simultaneously.

In the early 2000s, researchers introduced OETs, which combine optical stimulation with electric fields through the photoconductive effect of semiconductor materials [14]. This hybrid approach allows for the manipulation of particles using a relatively low optical power, making it particularly suitable for sensitive biological samples. OETs have demonstrated remarkable versatility in handling various types of particles, including cells and nanoparticles.

To overcome the fundamental diffraction limit of light, researchers have explored near-field optical technologies for particle manipulation. These include innovative approaches such as slot waveguides [15], which confine light to subwavelength dimensions by utilizing the strong evanescent field in a narrow air gap between two high-index dielectric materials. This unique configuration allows for intense light confinement, making slot waveguides particularly effective for enhancing optical forces and enabling the precise manipulation of nanoparticles and biomolecules in highly localized regions. Their compatibility with integrated photonic circuits further enhances their utility in lab-on-a-chip systems. Another groundbreaking approach involves photonic crystal resonators [16], which leverage periodic dielectric structures to create highly controlled optical fields. By engineering the bandgap properties of these structures, photonic crystal resonators can trap and manipulate particles with exceptional precision. The ability to tailor the resonance frequencies and field distributions of these devices makes them versatile tools for applications ranging from single-molecule studies to the assembly of photonic devices.

Additionally, photonic nanojets [17] – high-intensity, subwavelength light beams generated by illuminating dielectric microspheres – have emerged as a powerful tool for nanoscale manipulation. These nanojets exhibit a unique combination of high spatial resolution and extended depth of focus, enabling the precise trapping and manipulation of nanoparticles in three-dimensional space. Their simplicity and effectiveness have made them particularly attractive for applications in biosensing, nanofabrication, and single-cell analysis. Furthermore, nanofibers [18] have been employed to achieve highly localized optical fields through their ability to guide light with minimal loss and strong evanescent coupling. By tapering optical fibers to subwavelength diameters, researchers can create intense optical fields at the fiber surface, enabling the trapping and manipulation of nanoparticles, atoms, and even single molecules. Nanofibers are particularly valuable in quantum optics and biological applications, where their flexibility and compatibility with existing optical systems offer significant advantages. These near-field techniques have expanded the capabilities of optical manipulation into the realm of nanotechnology, opening new possibilities for single-molecule studies and nanofabrication applications.

Among these advancements, OETs, pioneered by Ming C. Wu and his team in 2005, have emerged as a promising technology, gaining considerable attention due to their exceptional performance across various domains [14]. This potent opto-electro-fluidic approach uses light-induced electrokinetic forces to manipulate and actuate objects at the nanoscale and microscale. By integrating the strengths of photonics and electronics, OETs offer an accessible solution for versatile and concurrent micromanipulation tasks. In contrast to OT, which requires coherent laser sources, OET systems typically employ a regular LED light source along with a digital micromirror device (DMD). This DMD modulates LED light to create desired light patterns, whether static or dynamic. These patterns are then projected onto the photoconductive substrate of an OET device through an optical microscope. Without light exposure, the photoconductive substrate acts as a high-impedance resistor. However, upon illumination, the photoelectric effect drastically reduces its impedance, effectively turning it into a conductor. As a result, "light-induced virtual electrodes" are established on the substrate, creating differentiated illuminated and dark regions. These virtual electrodes generate an uneven electric field within the liquid medium, interacting with samples to apply electrokinetic forces that manage their positions. Notably, the spatial distribution of this nonuniform electric field can be dynamically adjusted through the projected light patterns, enabling the flexible and swift manipulation of multiple targets at once. Compared to OTs, OETs allow simultaneous manipulation of up to 10,000 traps, demonstrating their capacity for massively parallel operations. Moreover, OETs can exert significantly stronger manipulation forces at a given light intensity. This technology proves suitable where OT might falter, such as in the simultaneous handling of many targets or moving objects across a wide size spectrum, ranging from tens of nanometers to hundreds of micrometers. Additionally, the operation of OETs does not necessitate a coherent light source and functions effectively at low light power densities (below 1 W/cm^2), making it optimal for working with heat-sensitive biological specimens, unlike other heat-assisted optical micromanipulation methods.

1.2 Importance and Applications in Various Fields

OETs stand out among optical micromanipulation techniques for their ability to deliver stronger manipulation forces at comparable light intensities. This unique advantage makes OETs particularly well-suited for the parallel manipulation of multiple objects, including micromachines with large, complex structures or composed of various components. The manipulation force in OETs is primarily governed by the interaction between the electric field and the objects, allowing the precise control over micromachines made from a wide range of materials, including insulators, semiconductors, and conductors. This versatility positions OETs as a powerful tool in advancing research on micromachinery, driving the development of sophisticated, optically controlled micromachines across diverse applications.

OETs have emerged as a versatile and powerful tool for the manipulation and assembly of various nanomaterials in non-biological applications. Employing the principles of light-induced dielectrophoresis (DEP), OETs allow precise control over nanomaterials, offering unique advantages for research and industry. Among the wide array of materials manipulated using OETs are semiconductor and metallic nanowires, carbon nanotubes, and graphene nanoplatelets. These materials play significant roles in the development of advanced electronic and photonic devices due to their exceptional electrical and thermal properties.

Furthermore, conductive nanoparticles and metal ions can also be intricately organized and positioned using OET. The manipulation capabilities of OET effectively cover nanoparticles ranging from just a few tens to several hundreds of nanometers, highlighting its versatility at the nanoscale. This ability facilitates the construction of complex nanoscale architectures and paves the way for innovations in nanotechnology and materials science.

In addition to nanomaterials, OETs are adept at handling microscale objects. These include not only dielectric and metallic micro-particles but also versatile media such as oil and water droplets, gas bubbles, and even specific electronic and photonic components. Such diverse applications demonstrate the proficiency of OETs in managing objects several microns to hundreds of microns in size, making it indispensable for the microsystem technology and lab-on-a-chip applications.

To ensure the effectiveness and efficiency of OET systems, polystyrene spherical microbeads are frequently utilized as standard test samples. The predictable shapes and favorable dielectric properties of these microbeads render them ideal for characterizing OET devices. By leveraging these microbeads, researchers can optimize the system performance and conduct in-depth studies of electrokinetic phenomena under varying experimental conditions. As a result, these microbeads are prevalent not only in device calibration but also in advancing the understanding of the underlying physics that governs OET operations.

In the biomedical arena, OETs have become an invaluable tool, offering novel and effective solutions for manipulating biological materials. Their application spans several cutting-edge areas, such as cell sorting, cell analysis, cell electroporation, and bio-fabrication, each of which plays a critical role in advancing biomedical research and therapeutic development.

A significant advantage of OETs over traditional OTs lies in the reduced reliance of the methodology on high-power light beams. Traditional OT systems often require intense laser light to trap and manipulate cells, which can inadvertently induce heating and cause potential physiological damage to the delicate biological samples. This issue poses a challenge, as it may compromise cell viability and affect the outcome of experiments, particularly in sensitive contexts.

The OET technology, on the other hand, minimizes these risks by employing the principles of light-induced DEP at much lower light intensities. Consequently, it provides a substantially gentler alternative for interacting with biological samples, avoiding the detrimental effects associated with optical heating. This characteristic makes OETs particularly advantageous for manipulating living cells, ensuring their integrity while enabling precise control over their positioning and sorting.

In cell sorting applications, OETs allow for the efficient separation of different cell types based on their unique dielectric properties. This capability is essential for isolating specific cell populations from heterogeneous samples in both research and clinical settings. The technology's suitability for cell analysis also facilitates the study of cellular behaviors and interactions under various conditions, contributing valuable insights into cellular biology and pathology.

Beyond manipulation and analysis, OETs offer significant potential in cell electroporation, where precise electrical fields are applied to temporarily permeabilize cell membranes. This process enables the introduction of foreign molecules, such as drugs or genetic material, into cells, proving instrumental for genetic engineering and drug delivery applications.

Furthermore, in the field of bio-fabrication, the precise control of OETs facilitates the construction and manipulation of cell-based biostructures and micromachines. This capability can lead to the development of complex tissue models and biomimetic devices, which have promising applications in tissue engineering and regenerative medicine.

Overall, OETs demonstrate substantial potential across a wide range of scientific and technological applications. Its versatility and efficacy in manipulating diverse materials highlight its transformative role in both non-biological and biomedical fields. As researchers continue to explore and refine the OET technology, its applications are likely to expand, paving the way for groundbreaking discoveries and innovations in multiple disciplines.

References

1 Giessibl, F.J. (2003). Advances in atomic force microscopy. *Reviews of Modern Physics* 75 (3): 949.

2 Gosse, C. and Croquette, V. (2002). Magnetic tweezers: micromanipulation and force measurement at the molecular level. *Biophysical Journal* 82 (6): 3314–3329.

3 Ozcelik, A., Rufo, J., Guo, F. et al. (2018). Acoustic tweezers for the life sciences. *Nature Methods* 15 (12): 1021–1028.

4 Ashkin, A., Dziedzic, J.M., Bjorkholm, J.E., and Chu, S. (1986). Observation of a single-beam gradient force optical trap for dielectric particles. *Optics Letters* 11 (5): 288–290.

5 Ashkin, A. (1970). Acceleration and trapping of particles by radiation pressure. *Physical Review Letters* 24 (4): 156.

6 Berkowitz, R. (2018). Father of optical trapping awarded a share of the Nobel Prize in Physics. *Physics Today* 71 (12): 14–17.

7 Lin, L., Wang, M., Peng, X. et al. (2018). Opto-thermoelectric nanotweezers. *Nature Photonics* 12 (4): 195–201.

8 Juan, M.L., Righini, M., and Quidant, R. (2011). Plasmon nano-optical tweezers. *Nature Photonics* 5 (6): 349–356.

9 Xie, Y., Zhao, C., Zhao, Y. et al. (2013). Optoacoustic tweezers: a programmable, localized cell concentrator based on opto-thermally generated, acoustically activated, surface bubbles. *Lab on a Chip* 13 (9): 1772–1779.

10 Constable, A., Kim, J., Mervis, J. et al. (1993). Demonstration of a fiber-optical light-force trap. *Optics Letters* 18 (21): 1867–1869.

11 Sasaki, K., Koshioka, M., Misawa, H. et al. (1991). Laser-scanning micromanipulation and spatial patterning of fine particles. *Japanese Journal of Applied Physics* 30 (5B): L907.

12 Svoboda, K. and Block, S.M. (1994). Biological applications of optical forces. *Annual Review of Biophysics and Biomolecular Structure* 23 (1): 247–285.

13 Padgett, M. and Di Leonardo, R. (2011). Holographic optical tweezers and their relevance to lab on chip devices. *Lab on a Chip* 11 (7): 1196–1205.

14 Chiou, P.Y., Ohta, A.T., and Wu, M.C. (2005). Massively parallel manipulation of single cells and microparticles using optical images. *Nature* 436 (7049): 370–372.

15 Yang, A.H., Moore, S.D., Schmidt, B.S. et al. (2009). Optical manipulation of nanoparticles and biomolecules in sub-wavelength slot waveguides. *Nature* 457 (7225): 71–75.

16 Mandal, S., Serey, X., and Erickson, D. (2010). Nanomanipulation using silicon photonic crystal resonators. *Nano Letters* 10 (1): 99–104.

17 Li, Y.-C., Xin, H.-B., Lei, H.-X. et al. (2016). Manipulation and detection of single nanoparticles and biomolecules by a photonic nanojet. *Light: Science & Applications* 5 (12): e16176.

18 Praveen Kamath, P., Sil, S., Truong, V.G., and Nic Chormaic, S. (2023). Particle trapping with optical nanofibers: a review. *Biomedical Optics Express* 14 (12): 6172–6189.

2

Fundamentals of Optoelectronic Tweezers

In the last 10 years, significant research efforts have been dedicated to enhancing the optoelectronic tweezer (OET) technology. This method employs electrokinetic forces that are created locally by casting light patterns onto optoelectronic semiconductor surfaces, enabling the trapping and manipulation of objects at the nano and microscale levels. These research efforts have led to the development of a versatile tool with a wide range of materials, structures, and functions for various applications and scientific investigations. In this discussion, we will explore the OET technology, focusing on its various device structures and the different mechanisms by which it operates.

2.1 Types of OET Device

OET devices can be categorized based on their operational principles into two main types. The first type is the conventional OET device, which operates with the aid of an external voltage source. The second type, referred to as a photovoltaic OET device, functions without the necessity of an external voltage source. More detailed information on both the conventional OET and photovoltaic OET devices will be provided in the next section [1].

2.1.1 Conventional OET Device

OET systems fundamentally operate on the basis of dynamically generating localized electric-field gradients using low-intensity light inputs. The field's orientation and generation process depend heavily on the device design, which is typically divided into two primary categories: vertical-field OET and lateral-field OET [2], as illustrated in Figure 2.1a and b, respectively. In a vertical-field OET, the electric field is set perpendicular to the surface of the device. This setup uses two glass substrates, each coated with a 300-nm indium tin oxide (ITO) layer. The bottom substrate additionally houses a 1-μm-thick layer of hydrogenated amorphous silicon (a-Si:H), applied through plasma-enhanced chemical vapor deposition (PECVD). The suspension containing the particles is situated between these substrates, separated by a thin layer of a double-sided tape. An electrical bias is applied across

Optoelectronic Tweezers for Manipulation at the Micro and Nano Scale, First Edition.
Lin Feng, Menglu Tan, Ao Wang, and Jiaying Zhang.
© 2026 WILEY-VCH GmbH. All rights reserved, including rights for text and data mining and training of artificial intelligence technologies or similar technologies. Published 2026 by WILEY-VCH GmbH.

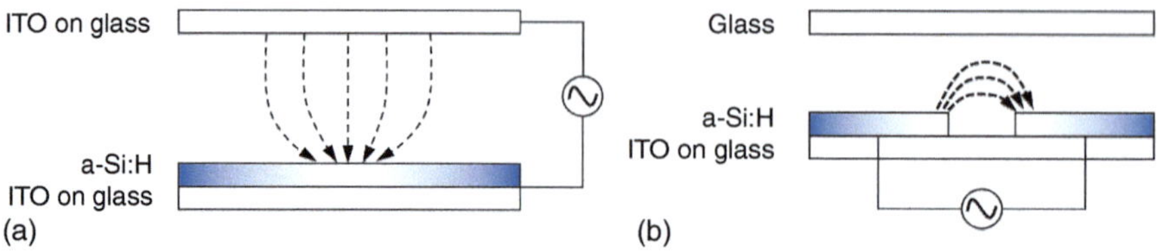

Figure 2.1 (a) Illustration of a vertical OET device. (b) Illustration of a lateral OET device. *Source*: Lin Feng.

the ITO layers to develop a perpendicular field. Initially, this field is confined mostly to the a-Si:H layer due to its high resistance. However, when illuminated, the light is absorbed, creating electron–hole pairs that significantly enhance the a-Si:H's conductivity. This process shifts the electric field to the liquid layer at the points of illumination, thereby producing localized field gradients that exert dielectrophoretic forces on particles in the vicinity [2].

Conversely, the lateral-field OET setup is characterized by a parallel orientation of the field relative to the substrate surface. Achieved through the use of interdigitated electrodes composed of ITO and a-Si:H, this setup necessitates an extra step where these electrodes are precisely patterned and etched. The solution carrying the particles is placed between the lateral-field OET substrate and a non-conducting top substrate. An external bias applied here keeps the field contained within the a-Si:H layer in the absence of light. Upon illumination, the conductivity of a-Si:H increases significantly, causing the field to shift into the solution between electrodes, forming field gradients that influence particle movement through DEP [2].

A notable benefit of lateral-field OET is its ability to manipulate multiple objects simultaneously. According to Ohta et al., the system has effectively demonstrated the precise positioning and angular control of four silver nanowires, each with a diameter of 80 nm [3].

Traditional OET systems perform poorly in high-conductivity environments like typical cell culture media and physiological buffers, which often have conductivities around 1.4 S/m, far exceeding the device's optimal range of 0.01–0.1 S/m. To work around this limitation, electrolytes in media must sometimes be replaced with non-electrolytes to sustain cell viability temporarily, yet this compromise affects normal cellular functions and viability over time, limiting OET's biological applications. To address these challenges, the newly developed phototransistor-based OET (Ph-OET) device is introduced, allowing effective manipulation in conditions like cell culture media where high conductivity is present (Figure 2.2a). By utilizing phototransistors that are over 500 times more conductive than the amorphous silicon used in traditional systems, the Ph-OET effectively operates with light intensities as low as 1 W/cm^2. This technology has been successfully applied to trap cells such as HeLa and Jurkat in widely used media, achieving significant transport speeds and forces. The Ph-OET device operates using a layered structure that includes a top ITO electrode, a liquid medium, a phototransistor array, and a bottom electrode, with an AC bias applied across them. In the absence of light, the phototransistor remains in an off state with high impedance, leading to a weak electric field in the liquid. But when illuminated, its impedance drops, creating

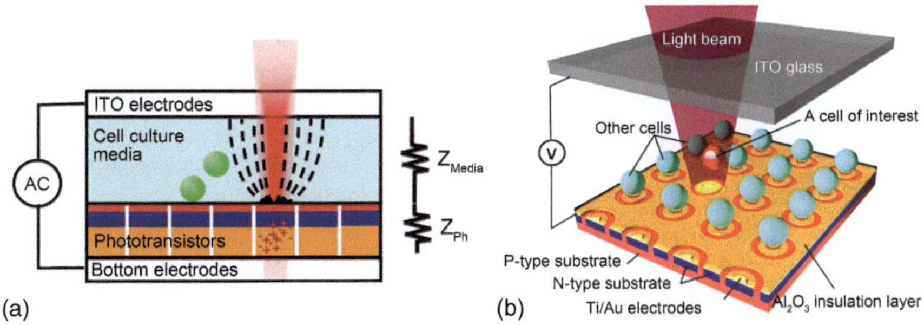

Figure 2.2 (a) Schematic of the Ph-OET device. (b) Schematic of a self-locking Ph-OET platform. *Source*: (a) Hsan-yin Hsu, 2010 / with permission of Royal Society of Chemistry. (b) Yajia Yang, 2016 / Springer Nature / CC BY 4.0.

a nonuniform electric field to manipulate cells via DEP. These innovations allow the Ph-OET to open up new possibilities in biological applications, such as drug screening and cell sorting, while retaining the benefits of standard OET systems [4].

Another variant of Ph-OET, developed by Chiou's group, enables the automatic trapping and release of cells in the dark and illuminated states, respectively (Figure 2.2b). A distinguishing characteristic is its ability to maintain a self-locking function across the entire active area when in a dark state. The strength of this locking effect is influenced by the voltage drop across the liquid layer in the dark state, which is adjusted based on factors such as the insulation layer's thickness and dielectric constant, the frequency of the AC current, and the conductivity of the medium. Ideally, during operation, the voltage drop in the dark state should approximate 50% of the total applied voltage. This configuration ensures that adequate dielectrophoretic forces can be generated to lock cells in place during the dark state, while still allowing for sufficient voltage variation in the illuminated state to release the cells [5].

While Ph-OET is pivotal for numerous applications, its complex fabrication process involving multiple microfabrication steps and semiconductor ion implantation renders it challenging and costly to produce, limiting accessibility for most research laboratories. Despite the cost-effectiveness and ease of fabrication of organic photoconductive materials, their drawbacks in terms of surface roughness and photo/water stability underscore the continued preference for a-Si:H as the predominant photoconductive material in OET research, given its material stability, robustness, simplicity of fabrication, and cost-effectiveness [1].

2.1.2 Photovoltaic OET Device

Photovoltaic OET devices represent a significant departure from conventional OET systems by achieving autonomous operation without the necessity for an external power supply. These devices explore the bulk photovoltaic effect, where photorefractive crystal materials facilitate the directional movement of charge carriers when exposed to light [6]. This process is aligned with the polar axis of the crystal, resulting in the generation of a space-charge field. The electric field thus produced is highly nonuniform across the crystal's surface and serves as the propulsion mechanism for

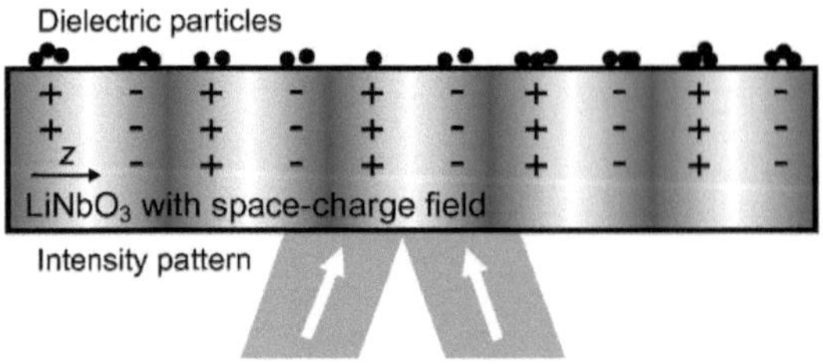

Figure 2.3 Schematic of a photovoltaic OET setup. *Source:* H. A. Eggert 2007/ with permission of AIP Publishing.

photovoltaic OETs, as illustrated in Figure 2.3. Among the various materials that exhibit this effect, $LiNbO_3$, particularly when doped with impurities like Fe and Cu, is preferred for its exceptional ability to develop strong internal electric fields, which can reach several kilovolts per millimeter [7]. Such materials are fundamental in the effective deployment of $LiNbO_3$-based photovoltaic OET systems, which have been adapted for patterning and assembling nano- and micro-scale materials [8].

One of the standout advantages of photovoltaic OETs over the standard systems is their ability to autonomously generate a space–charge electric field. However, directing and localizing electrical carriers to create this field under low light power densities (generally ranging from 1 to 10 mW/cm^2) is a time-consuming process, taking several minutes [9]. This prolonged duration can hinder the system's temporal resolution, posing challenges for achieving quick micromanipulation with minimal light intensity. For the real-time manipulation of particles and droplets, photovoltaic OET systems demand much higher light power densities, typically between 1 and 100 W/cm^2. These requirements can create additional complications, particularly when dealing with samples that are sensitive to heat and light. As such, further advancements in photorefractive crystal materials are anticipated – materials that would operate efficiently under low light conditions and exhibit a faster response time in forming spontaneous electric fields. These enhancements could substantially improve the efficacy and adaptability of photovoltaic OETs to real-time micromanipulation applications, potentially broadening their scope and utility in a range of scientific and industrial fields.

2.2 Working Mechanisms

Figure 2.4 depicts the structure of the OET device employed in our experiments. The liquid containing the target cells or particles is placed between two surfaces: the top is a transparent glass coated with a conductive layer of ITO, and the bottom is a photosensitive layer composed of multiple layers without distinct features. These bottom layers include an ITO-coated glass, a heavily doped $n+$ layer of a-Si:H, followed by an undoped a-Si:H layer, and capped with a silicon nitride layer. To operate, an AC signal with a peak-to-peak voltage of 10 V is applied. When light from an illumination source, such as a light-emitting diode (LED) with a wavelength of 625 nm, is projected onto the photoconductive layer, it activates virtual electrodes on the surface. This results in the formation of nonuniform electric fields that facilitate the manipulation of particles using dielectrophoretic

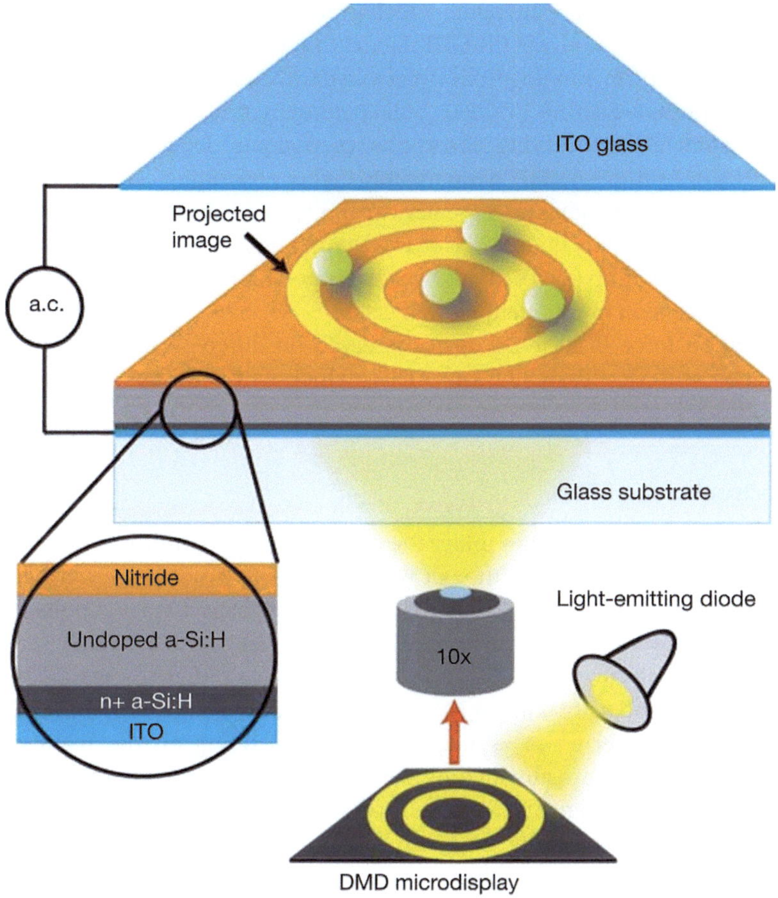

Figure 2.4 Schematic of OET. *Source*: Pei Yu Chiou 2005 / with permission of Springer Nature.

forces. The fabrication of these featureless layers does not require photolithography, making the device cost-effective and suitable for disposable applications. The OET-based optical manipulation system has two operational modes driven by the DEP forces: positive OET and negative OET. Depending on the frequency of the AC electric field and the dielectric properties of the particles' surfaces and interiors, particles may either be drawn to or repelled from the illuminated regions [9].

The AC voltage applied to an OET device is typically delivered through an external power source and can take various waveform shapes, such as square, sine, or triangle waves. Research indicates that a square wave, despite having the same peak-to-peak voltage and frequency as other waveforms, is particularly effective at generating strong manipulation forces. This strength in manipulation makes square waves the preferred choice for enhancing the OET device performance [10]. The frequency range for the AC bias generally extends from a few kilohertz to several hundred kilohertz, depending on the dielectric properties and conductivity of both the fluid medium and the particles involved. The choice of frequency influences the effectiveness of the electric field generated, which in turn affects the manipulation of the samples [11]. The sensitivity of the OET device's photoconductive materials to light spans a broad spectrum, ranging from violet to red wavelengths. Materials such as a-Si:H, titanyl phthalocyanine (TiOPc), bulk heterojunction (BHJ) polymers, and phototransistors exhibit varying absorption coefficients depending on the wavelength [12]. For example, a-Si:H has a notably higher absorption coefficient in the violet–blue region. This indicates a greater capability to produce electron–hole pairs and increase conductivity under shorter-wavelength light [13]. As a result, the enhanced electric field within an OET device leads to stronger manipulation forces, making it more effective in micromanipulation tasks.

The manipulation force in an OET device is based on different electrokinetic mechanisms such as DEP, EP, ACEO, and ACET, which will be discussed in the following sections [1].

2.2.1 Dielectrophoretic Effect

DEP is the motion of neutral particles under the influence of a nonuniform electric field, as shown in Figure 2.5. Even though these particles do not possess a net electrical charge, they can become polarized when an electric field is applied. This polarization results in the particle experiencing a net force, provided the electric field is non-uniform. The DEP force can be directed either toward areas of higher electric field intensity, known as positive DEP, or toward areas of lower electric field intensity, referred to as negative DEP, depending on the relative permittivity of the particle and surrounding medium. The fundamental principle of DEP can be described mathematically. The time-averaged DEP force acting on a spherical particle can be expressed by the formula [14]:

$$F_{\mathrm{DEP}} = 2\pi r^3 \varepsilon_m \mathrm{Re}[\mathrm{CM}] \nabla |E|^2 \tag{2.1}$$

where r is the radius of the spherical particle, ε_m is the permittivity of the medium, Re[CM] is the real part of Clausius–Mossotti factor, and E is the electric field magnitude. Clausius–Mossotti factor is given by [14]:

$$\mathrm{CM} = \frac{\varepsilon'_p - \varepsilon'_m}{\varepsilon'_p + \varepsilon'_m} \tag{2.2}$$

where ε'_p and ε'_m are the complex permittivity (CM) of the particle and the medium, respectively, which are expressed as:

$$\varepsilon'_p = \varepsilon_p - j\frac{\sigma_p}{\omega} \tag{2.3}$$

$$\varepsilon'_m = \varepsilon_m - j\frac{\sigma_m}{\omega} \tag{2.4}$$

where ε_p and ε_m are the permittivity of the particle and the medium, respectively; σ_p and σ_m are the conductivity of the particle and medium, respectively; and ω is the angular frequency. The CM serves as a metric of polarizability, dictating the polarity of the DEP force. When $\mathrm{Re[CM]} > 0$, indicating positive DEP, particles exhibit higher polarizability than the surrounding medium and are drawn toward

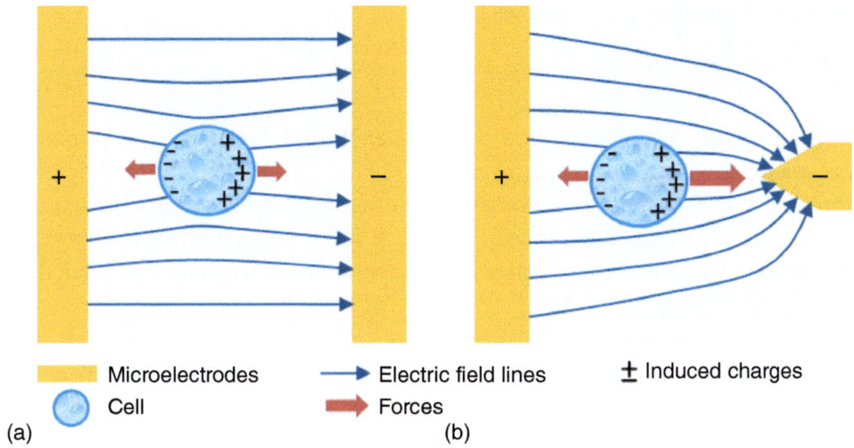

Figure 2.5 Diagram of the positive DEP. *Source*: Khashayar Khoshmanesh, 2011 / with permission of Elsevier.

the illuminated area with a stronger electric field compared to the dark region. Conversely, when Re[CM] < 0, a condition known as negative DEP arises. Under this condition, particles exhibit lower polarizability than the surrounding medium, causing them to be repelled from regions with higher electric field intensity, such as the illuminated areas.

While Eq. (2.1) serves as a useful conceptual framework for understanding the impact of various experimental conditions on the DEP force, it is constrained by several assumptions that could limit its applicability in practical scenarios. Specifically, the equation assumes that the size of the particle is considerably smaller than the gradients in the electric field, allowing for a uniform application of the DEP force across the particle's surface. This assumption implies that the particle behaves as if the electric field is constant over its volume, a condition that does not hold true in many experimental setups. In practice, these assumptions often do not align with experimental realities, resulting in discrepancies between the predicted and observed outcomes. This is because, in many experiments, particles can be comparable in size to the variations in the electric field, leading to nonuniform forces across the particle. Consequently, predictions based solely on Eq. (2.1) may be oversimplified and less accurate. For a more rigorous and precise determination of the DEP force acting on particles, it is beneficial to adopt a model grounded in Lorentz law forces, employing the Maxwell stress tensor, which can be expressed as [15]:

$$\sigma_{ij} = \varepsilon_0 E_i E_j + \frac{1}{\mu_0} B_i B_j - \frac{1}{2}\left(\varepsilon_0 |E|^2 + \frac{1}{\mu_0}|B|^2\right)\delta_{ij} \tag{2.5}$$

where σ_{ij} is the ij^{th} element of the second rank Maxwell stress tensor; ε_0 and μ_0 are the vacuum permittivity and permeability, respectively; E and B are the electric and magnetic fields, respectively; and δ_{ij} is Kronecker's delta. Through the integration of the Maxwell stress tensor across the surface area of the volume exposed to the electromagnetic field, the manipulation force acting on a dielectric object within this field can be determined as follows [15]:

$$F_{\mathrm{DEP}} = \int_{S} \sigma \cdot n dS \tag{2.6}$$

where n is the unit vector and S is the surface enclosing the bead. This approach allows for a detailed analysis that considers the spatial variation of the electric field and the resulting distribution of forces on the particle. By accounting for these complexities, researchers can achieve a more accurate prediction of the particle behavior under DEP, leading to better experimental alignment and more reliable application outcomes.

2.2.2 Electrophoretic Effect

The electrophoretic effect refers to the motion of charged particles or molecules in a fluid under the influence of an electric field. When an electric field is applied, charged entities in a solution or colloid will migrate toward the electrode of opposite charge. This movement is driven by the force exerted by the electric field on the particle's charge and balanced by the frictional force from the fluid medium.

The movement of charged micro-objects in relation to a fluid under the influence of an electric field at DC or low frequency (i.e. < 10 Hz) is known as electrophoresis (EP), as shown in Figure 2.6. This phenomenon stems from the Coulomb force, which is given by [16]:

$$F_{\mathrm{Coulomb}} = Eq \tag{2.7}$$

where E is the electric field strength, and q is the charge of the particle.

Typically, cells and various biological molecules carry a net negative charge. This characteristic serves as a fundamental principle for techniques such as capillary and gel electrophoresis, where charged molecules are separated based on their migration in response to an electric field. For example, in gel electrophoresis, negatively charged DNA segments move through a gel matrix toward a positive electrode, enabling the separation and analysis of DNA fragments according to size.

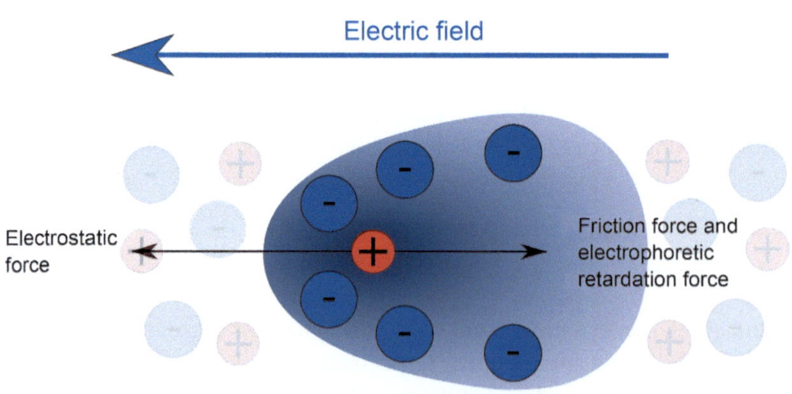

Figure 2.6 Diagram of the electrophoresis effect. *Source*: Daniele Pugliesi / Wikimedia Commons / CC BY-SA 4.0 / https://en.wikipedia.org/wiki/Electrophoresis#/media/File:Motion_by_electrophoresis_of_a_charged_particle.svg / last accessed by October 03, 2025.

In the context of OETs, the electrophoretic force comes into play primarily when the system is subject to certain types of electrical signals. Specifically, electrophoretic considerations are significant when a DC or an AC signal with very low frequencies, typically below 10 Hz, is applied. At these low frequencies, the electric field has enough time to influence the charged particles or cells, causing them to migrate.

2.2.3 Electroosmotic Effect

The electroosmotic effect is a fluid dynamics phenomenon, where an applied electric field induces liquid flow through porous materials or along microchannels with charged surfaces. This process is driven by the interaction between the electric field and the electrical double layer (EDL) that forms at the interface between the liquid and the solid surface. When an electric field is applied, the mobile ions within the diffuse layer of the EDL migrate toward the electrode with the opposite charge. As these ions move, they drag the surrounding liquid with them, creating net fluid flow (Figure 2.7). When this process occurs in the presence of an AC electric field, it is specifically termed ACEO. In ACEO, the oscillating electric field induces ion movement in a manner that leads to a fluid flow along the surface, enabling the efficient manipulation of liquids at the microscale.

The electroosmotic flow velocity can be described by the Smoluchowski equation [17]:

$$v_{\mathrm{SLIP}} = -\frac{\varepsilon_m \zeta E_t}{\eta} \tag{2.8}$$

where ε_m is the permittivity of the liquid, ζ is the potential difference across the EDL, E_t is the tangential electric field strength, and η is the dynamic viscosity of the liquid. The ideal frequency range for the ACEO flow typically spans from several hundred Hz to several kHz, and this phenomenon can be characterized as [17]:

$$f_{\mathrm{OPT}} = \frac{\sigma_m \lambda_d}{2\pi \varepsilon_m D} \tag{2.9}$$

where σ_m is the conductivity of the liquid, λ_d is the thickness of EDL, and D is the thickness of the liquid chamber. In OET, the ACEO flow can be manipulated by the light-induced electric field to propel micro-objects toward the center of illumination. As the magnitude of the ACEO flow diminishes vertically from the OET bottom plate, micro-objects tend to aggregate and accumulate at the center of

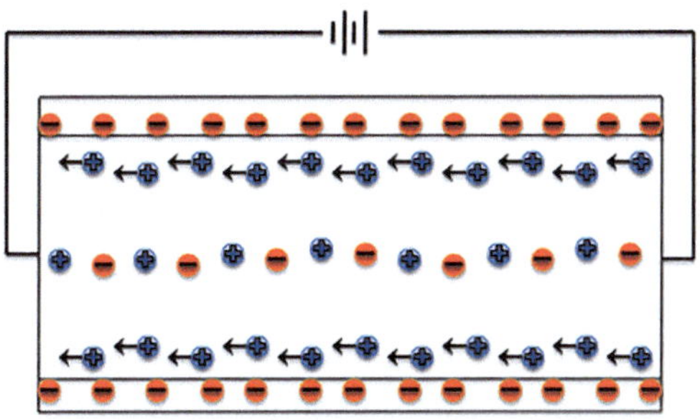

Figure 2.7 Diagram of the electroosmotic flow. *Source*: DWRosenow / Wikimedia Commons / CC BY-SA 4.0 / https://en.wikipedia.org/wiki/Electro-osmosis#/media/File:Electroosmotic_Flow_Depiction.tiff / last accessed by October 03, 2025.

illumination on the OET bottom plate. Studies have shown that ACEO is capable of displacing objects located at a considerable distance from the illuminated area (several millimeters away) [18]. Furthermore, ACEO exhibits enhanced efficacy in moving smaller objects [19]. Consequently, ACEO can be effectively employed for particle sorting by size and for the manipulation of objects that are challenging to control using optically induced dielectrophoretic (ODEP) forces, such as nanoparticles and biomolecules.

2.2.4 Electrothermal Effect

The electrothermal effect is a phenomenon where an electric field induces temperature gradients in a conductive fluid, leading to fluid motion due to the resultant buoyancy and density differences. This effect can occur when AC or DC fields cause localized heating within the fluid, primarily due to resistive (Joule) heating. The electric field interacts with the temperature-dependent properties of the fluid, such as permittivity and conductivity, causing variations that drive fluid flow (Figure 2.8a). In microfluidic systems, the electrothermal effect is harnessed to manipulate fluids and particles without mechanical components (Figure 2.8b). The induced temperature gradients create forces within the fluid, which can be used for mixing, pumping, or concentrating particles. When a fluid exhibiting gradients in permittivity and conductivity is exposed to an AC electric field, it experiences fluid movement driven by a net body force [20]. This flow phenomenon is known as the ACET flow and can be mathematically described by the following equation [21]:

$$F_{\mathrm{ET}} = \frac{1}{2}\mathrm{Re}\left[\frac{\sigma_m \varepsilon_m}{\sigma_m + i\omega\varepsilon_m}(\kappa_\varepsilon - \kappa_\sigma)(\nabla T \cdot E)E^*\right] - \frac{1}{2}|E|^2 \kappa_\varepsilon \kappa_\sigma \nabla T \tag{2.10}$$

where σ_m is the conductivity of the medium, ε_m is the permittivity of the medium, ω is the angular frequency of the AC electric field, T is the thermal field strength, E is the electric field strength, E^* is the complex conjugate of the electric field, and κ_σ and κ_ε are constants, which represent the variations of the conductivity and permittivity per-unit temperature, respectively.

In OET, the ACET effect, in particular, is influenced by variations in fluid properties like conductivity and permittivity with temperature. These variations are constant factors that determine how ACET contributes to particle manipulation. The magnitude of the ACET effect is closely tied to the optical power density, which establishes the temperature gradient's strength. However, under typical low-power light conditions (less than $100\,\mathrm{W/cm^2}$) in OET, the impact of ACET is generally limited. Beyond manipulation forces, unintentional factors like nonspecific surface adhesion and electrostatic interactions between particles can hinder OET's effectiveness. To counteract these limitations, strategies such as employing chemical surfactants, antifouling coatings, lipid bilayers, and innovative device designs have been developed. These solutions aim to reduce interference and optimize the manipulation efficiency. OET's versatility and adaptability stem from its ability to fine-tune manipulation forces. By adjusting parameters such as optical power density, the distribution of light patterns, the conductivity of the medium, and the

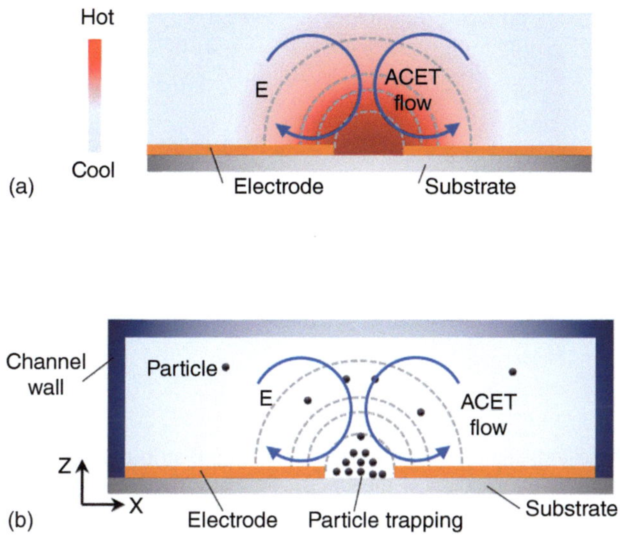

Figure 2.8 Diagram of the AC electrothermal flow. (a) Temperature gradients induced by an external electric field lead to fluid motion; (b) particle trapping under ACET. *Source*: Ahmed Abdelghany, 2024 / with permission of Elsevier.

frequency and amplitude of the electric field, users can customize the system to target specific particles effectively. This flexibility makes OET a powerful tool for a wide range of applications, allowing the precise control in various experimental settings.

2.2.5 Finite Element Simulation Analysis

To investigate the specific characteristics of the spatial distribution of dielectrophoretic forces, as well as the impact of factors such as the dielectric constant and conductivity of particles and solution, the frequency and magnitude of the applied electric field, and the radius of the particles on dielectrophoretic forces, the dielectrophoretic force distribution in the optoelectronic chip is simulated and analyzed using COMSOL Multiphysics 5.5a. The virtual electrode for illumination is set as a ring, with the illuminated area being conductive and the dark area being insulating. The electrode boundary satisfies the Dirichlet boundary condition: $V = V0$, while the remaining boundaries satisfy the Neumann boundary condition: $\partial v/\partial n = 0$. A sine signal with a frequency of 10 kHz and a voltage of $10\,V_{pp}$ is applied, with a liquid conductivity of 78 mS/m [22].

The simulation results are shown in Figure 2.9a and b. The simulation indicates that a peak electric field is formed near the boundary between light and dark areas. The OET force model is composed of a weak region at the non-electrode area and a strong electric field region at the electrode area, thus enabling the manipulation of dielectric particles. At the boundary between the bright and dark areas, dielectrophoretic forces are generated. The direction of the dielectrophoretic force can be either toward the region with high electric field density, i.e. the bright area, or toward

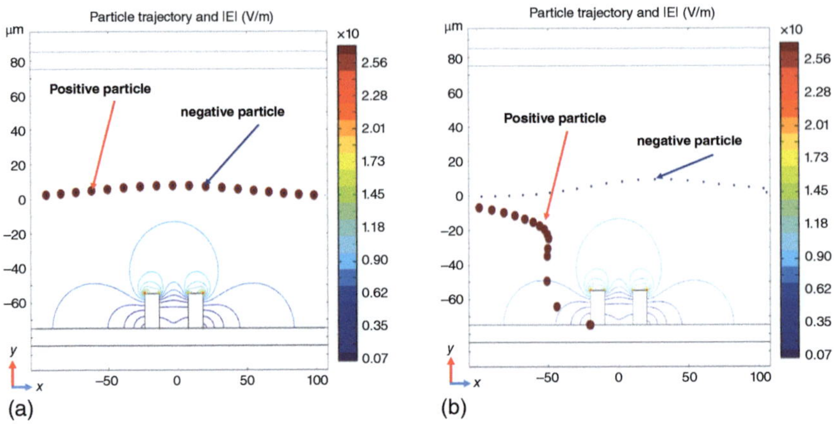

Figure 2.9 Distribution of photo-induced nonuniform electric field. (a) Distribution of potential and electric field direction; (b) intensity of the electric field at different lengths. *Source*: Shuzhang Liang, 2021 / MDPI / CC BY 4.0.

the region with low electric field density, i.e. the dark area. In addition to the polarization properties of the medium and particles, the frequency of the applied electric field also affects the polarity of the dielectrophoretic force. At different frequencies, the direction of the dielectrophoretic force varies. Furthermore, the capture of particles by the circular ring-shaped illumination pattern is not absolute. If the viscous force exerted on the particles by the liquid exceeds the maximum dielectrophoretic force, the particles will detach from the illumination pattern.

At the same time, the interaction between two particles with different dielectric properties was simulated and analyzed using electric field modeling. In the experiment, deionized water was used as the manipulating medium, with dielectric constant and conductivity values of 80 and 2×10^{-4} S/m, respectively. Regarding the dielectric properties of microorganisms, most biological particles have complex heterogeneous structures such as cell membranes, cytoplasm, and cell nuclei, each with different dielectric properties. Therefore, a multi-shell spherical model was used for simulation analysis, and the multi-shell model can be equivalent to a single-shell model. For the single-shell model of microorganisms, the cell membrane is typically composed of nonconductive lipids and proteins, making it a poor conductor. Therefore, the dielectric constant of the cell membrane is around 6, and the conductivity is approximately 2.5×10^{-7} S/m. The cytoplasm is a physiological saline solution, with proteins accounting for only a small fraction of its volume. Due to the presence of many ions and charged particles in the cell, the conductivity of the cytoplasm is higher than that of the cell membrane. Hence, the dielectric constant of the cytoplasm is about 60, and the conductivity is around 0.2 S/m.

As shown in Figure 2.10a, the electric field distribution between two particles before contact is displayed. After the two particles come into contact, the electric field distributions of the two particles interact, as shown in Figure 2.10b, where the red and blue circles represent the attraction of positive dielectric particles to the light spot and the repulsion of negative dielectric particles, respectively. The force acting

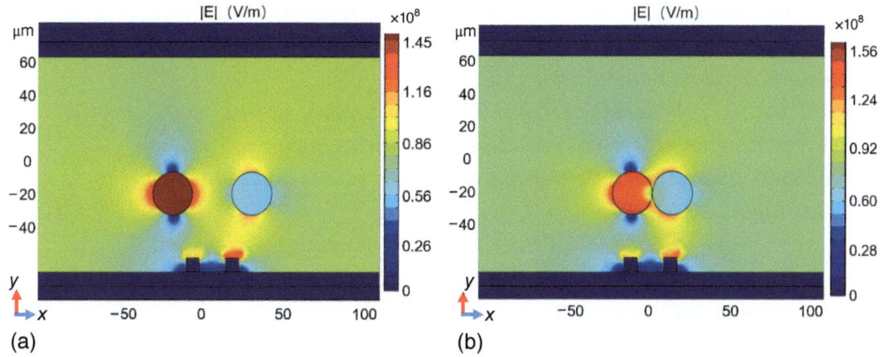

Figure 2.10 Distribution of electric field interactions between dielectric particles with different characteristics. (a) Distribution of electric field forces on positively charged microparticles and negatively charged microparticles before adhesion; (b) distribution of electric field forces acting on particles after adhesion, with red circles representing positive particles and blue circles representing negative particles. *Source*: Lin Feng.

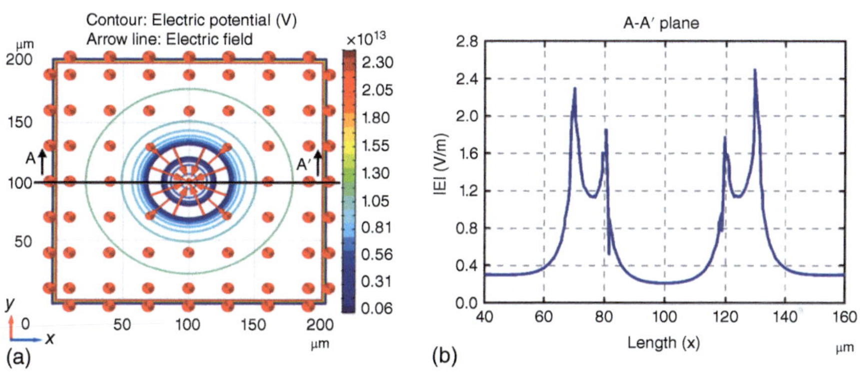

Figure 2.11 Force separation of particles with different dielectric properties under an electric field. (a) Particle separation situation without applied electric field; (b) particle separation situation with applied electric field. *Source*: Lin Feng.

on the two particles in the OET field is analyzed in Figure 2.11a and b. To verify the different effects of the dielectrophoretic force polarity on different particles, they were released into a nonuniform electric field, where the trajectories of the particles reflect the attraction and repulsion effects of the light spot on positive and negative particles, respectively. The simulation results provide a parameter basis for handling particles with different dielectric properties.

References

1 Zhang, S., Xu, B., Elsayed, M. et al. (2022). Optoelectronic tweezers: a versatile toolbox for nano-/micro-manipulation. *Chemical Society Reviews* 51 (22): 9203–9242.

2 Valley, J.K., Ohta, A.T., Hsu, H.-Y. et al. (2009). Optoelectronic tweezers as a tool for parallel single-cell manipulation and stimulation. *IEEE Transactions on Biomedical Circuits and Systems* 3 (6): 424–431.

3 A. T. Ohta, S. L. Neale, H.-Y. Hsu, J. K. Valley, and M. C. Wu. (2008). Parallel assembly of nanowires using lateral-field optoelectronic tweezers. *2008 IEEE/LEOS International Conference on Optical MEMs and Nanophotonics* (11–14 August 2008). IEEE, pp. 7–8.

4 Hsu, H.Y., Ohta, A.T., Chiou, P.Y. et al. (2010). Phototransistor-based optoelectronic tweezers for dynamic cell manipulation in cell culture media. *Lab on a Chip* 10 (2): 165–172.

5 Yang, Y., Mao, Y., Shin, K.-S. et al. (2016). Self-locking optoelectronic tweezers for single-cell and microparticle manipulation across a large area in high conductivity media. *Scientific Reports* 6 (1): 22630.

6 Eggert, H., Kuhnert, F., Buse, K. et al. (2007). Trapping of dielectric particles with light-induced space-charge fields. *Applied Physics Letters* 90 (24).

7 Villarroel, J., Burgos, H., García-Cabañes, Á. et al. (2011). Photovoltaic versus optical tweezers. *Optics Express* 19 (24): 24320–24330.

8 Esseling, M., Zaltron, A., Sada, C., and Denz, C. (2013). Charge sensor and particle trap based on z-cut lithium niobate. *Applied Physics Letters* 103 (6).

9 Chiou, P.Y., Ohta, A.T., and Wu, M.C. (2005). Massively parallel manipulation of single cells and microparticles using optical images. *Nature* 436 (7049): 370–372.

10 W. Liang, S. Wang, Z. Dong,G.-B. Lee, and W. J. Li, "Optical spectrum and electric field waveform dependent optically-induced dielectrophoretic (ODEP) micro-manipulation," *Micromachines*, vol. 3, no. 2, pp. 492–508, 2012.

11 Zhu, X., Yi, H., and Ni, Z. (2010). Frequency-dependent behaviors of individual microscopic particles in an optically induced dielectrophoresis device. *Biomicrofluidics* 4 (1).

12 Wang, W., Lin, Y.-H., Wen, T.-C. et al. (2010). Selective manipulation of microparticles using polymer-based optically induced dielectrophoretic devices. *Applied Physics Letters* 96 (11).

13 Valley, J.K., Jamshidi, A., Ohta, A.T. et al. (2008). Operational regimes and physics present in optoelectronic tweezers. *Journal of Microelectromechanical Systems* 17 (2): 342–350.

14 Pethig, R. (2010). Dielectrophoresis: status of the theory, technology, and applications. *Biomicrofluidics* 4 (2).

15 Wang, X., Wang, X.B., and Gascoyne, P.R.C. (1997). General expressions for dielectrophoretic force and electrorotational torque derived using the Maxwell stress tensor method. *Journal of Electrostatics* 39 (4): 277–295.

16 Hwang, H. and Park, J.-K. (2011). Optoelectrofluidic platforms for chemistry and biology. *Lab on a Chip* 11 (1): 33–47.

17 Chiou, P.-Y., Ohta, A.T., Jamshidi, A. et al. (2008). Light-actuated AC electroosmosis for nanoparticle manipulation. *Journal of Microelectromechanical Systems* 17 (3): 525–531.

18 Jamshidi, A., Neale, S.L., Yu, K. et al. (2009). NanoPen: dynamic, low-power, and light-actuated patterning of nanoparticles. *Nano Letters* 9 (8): 2921–2925.

19 Liang, W., Liu, L., Lai, S.H.-S. et al. (2014). Rapid assembly of gold nanoparticle-based microstructures using optically-induced electrokinetics. *Optical Materials Express* 4 (11): 2368–2380.

20 Kumar, A., Kwon, J.-S., Williams, S.J. et al. (2010). Optically modulated electrokinetic manipulation and concentration of colloidal particles near an electrode surface. *Langmuir* 26 (7): 5262–5272.

21 Castellanos, A., Ramos, A., Gonzalez, A. et al. (2003). Electrohydrodynamics and dielectrophoresis in microsystems: scaling laws. *Journal of Physics D: Applied Physics* 36 (20): 2584.

22 Liang, S.Z. (2020). *Based study on Microorganism Robot Control via Multifunctional Photoelectronic Tweezers System*. Beijing, China: Beihang University.

3

Optics and Light Source

Grasping the foundational optical concepts is vital for optimizing OET's performance in various applications. This chapter begins by covering the essential optical principles of OETs, highlighting the impact of lens imaging and optical aberrations. Furthermore, it discusses the critical role of various light sources in OETs, examining their principles and characteristics in research contexts.

3.1 Optics Involved in Optoelectronic Tweezers

3.1.1 Principle of Lens Imaging

A lens is an optical component that allows the passage of light and alters its path by either concentrating or spreading the light beam through refraction. Lenses form the cornerstone of various optical devices, ranging from the simplest forms like magnifying glasses, which are used to enlarge the appearance of objects, to the highly complex systems found in cameras, microscopes, telescopes, and eyeglasses. Each of these devices utilizes lenses to manipulate light so as to enhance human vision or capture detailed images of the microscopic or distant world.

Single-element lenses, commonly referred to as simple lenses, are designed with a single optical component that refracts light to focus or scatter it. These lenses are often used in straightforward applications where minimal correction for optical aberrations is needed, such as magnifying glasses and basic reading glasses. In contrast, compound lenses, also known as complex lenses, are composed of multiple lens elements or groups of elements working together within an optical system. This configuration is employed to correct for various optical imperfections and to enhance the image quality. By carefully aligning and shaping the individual components, compound lenses achieve superior performance in applications requiring precise light manipulation, such as high-quality cameras, microscopes, and telescopes, where sharpness, clarity, and focus are critical. Lenses find applications in various imaging instruments such as telescopes, binoculars, and cameras. They are also utilized as visual aids in eyeglasses to rectify vision impairments like myopia and hypermetropia.

Lenses are broadly categorized into two primary types: convex and concave, also referred to as converging and diverging lenses, respectively. Convex lenses have

Optoelectronic Tweezers for Manipulation at the Micro and Nano Scale, First Edition.
Lin Feng, Menglu Tan, Ao Wang, and Jiaying Zhang.
© 2026 WILEY-VCH GmbH. All rights reserved, including rights for text and data mining and training of artificial intelligence technologies or similar technologies. Published 2026 by WILEY-VCH GmbH.

outward-curving surfaces that enable them to focus incoming light rays to a single point. This property makes convex lenses suitable for applications where light concentration is necessary, such as in hyperopia-correcting eyewear or in devices like cameras and projectors that require precise image focus.

On the other hand, concave lenses possess inward-curving surfaces that cause light rays to spread out. As a result, these lenses are effective in addressing myopia (nearsightedness) and are often used in conjunction with other optical components to manage the direction and dispersion of light within complex optical systems.

Lenses can also be further divided based on the curvature of their two optical surfaces (Figure 3.1). Biconvex lenses, or simply convex lenses, have two convex surfaces. If these surfaces have identical curvature radii, the lens is termed equiconvex. Conversely, biconcave lenses, or concave lenses, have two inward-curving surfaces. In cases where one side is flat, the lens is classified as plano-convex or plano-concave, determined by the curvature of the other surface. A lens comprising one convex and one concave side is known as a convex–concave lens or meniscus lens, often used in corrective lenses due to their ability to reduce certain optical aberrations.

When a collimated light beam passes through a biconvex or plano-convex lens situated in a medium of lower refractive index, the light converges to a focal point located behind the lens (Figure 3.2). These lenses are classified as positive or converging lenses. In the context of a thin lens in air, the focal length, represented as "f," is the distance from the lens to this focal point. An extended hemispherical lens represents a specialized type of plano-convex lens, characterized by a fully hemispherical curved surface and a thickness that exceeds the radius of curvature, offering unique focusing capabilities for particular optical applications.

A ball lens represents an extreme example of a thick convex lens, characterized by its completely round shape. In novelty photography, it is frequently referred to as a "lensball." The spherical design of a ball lens offers the advantage of being omnidirectional. However, for the majority of optical glass materials, the focal point of a ball lens is situated near the surface of the ball. Due to the pronounced curvature of the ball in relation to its size, optical aberrations are significantly more pronounced compared to thin lenses, with the exception of chromatic aberration.

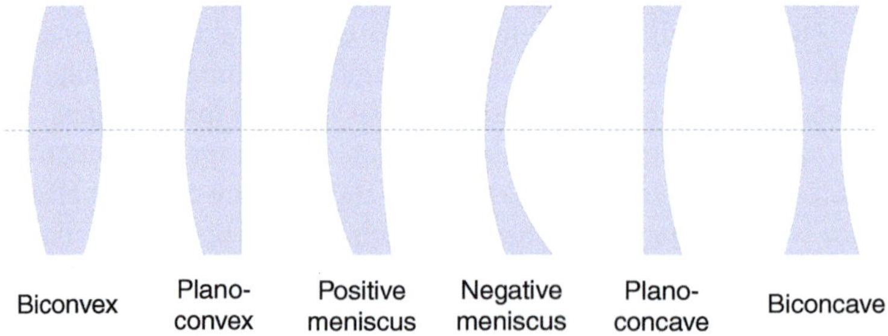

Figure 3.1 Types of lenses. *Source*: ElfQrin / Wikimedia Commons / CC BY-SA 4.0 / https://commons.wikimedia.org/wiki/File:Lenses_en.svg / last accessed by October 03, 2025.

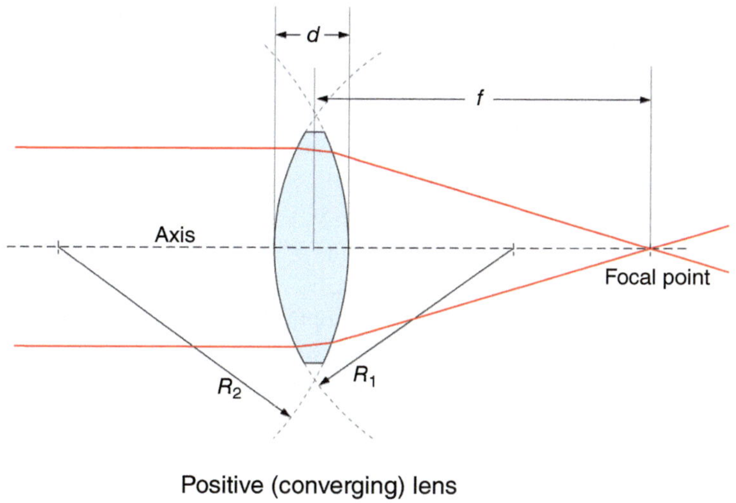

Figure 3.2 Positive lens. *Source*: DrBob / Wikimedia Commons / CC BY-SA 4.0 / https://commons.wikimedia.org/wiki/File:Lens1.svg / last accessed by October 03, 2025.

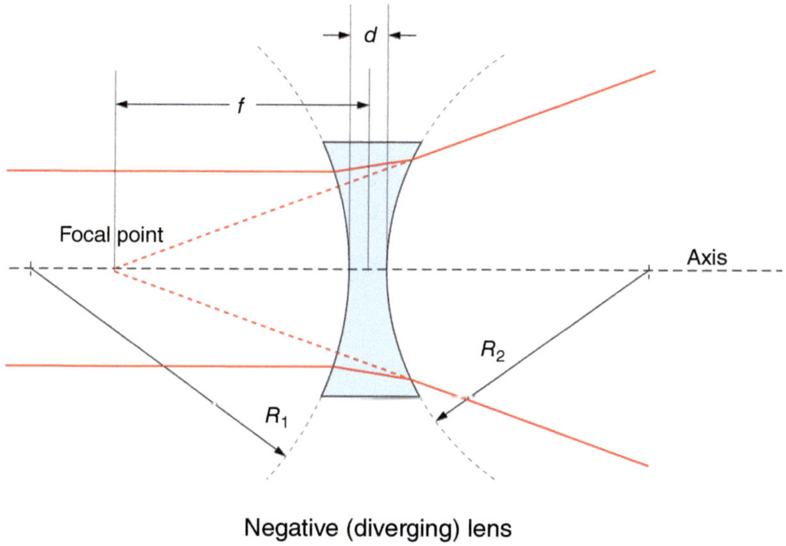

Figure 3.3 Negative lens. *Source*: Dr Bob / Wikimedia Commons / CC BY-SA 4.0 / https://commons.wikimedia.org/wiki/File:Lens1b.svg / last accessed by October 03, 2025.

A diverging lens, also known as a concave lens, is designed to spread out light rays that pass through it (Figure 3.3). These lenses have at least one inwardly curved surface, and their principal function is to diverge incoming parallel light rays away from a common point. When light rays enter a diverging lens, they are refracted as they pass through the lens material. Due to the lens's shape and the difference in refractive index between the lens material and the surrounding medium (usually air), the rays bend outward. This bending creates the effect that the light rays are emanating from a single point on the same side of the lens as the incoming rays. This point is known as the virtual focal point since the rays do not physically meet but appear to diverge from this location.

The (effective) focal length f positioned in air or vacuum, as it pertains to paraxial rays, can be determined using the lensmaker's equation [1]:

$$\frac{1}{f} = (n-1)\left[\frac{1}{R_1} - \frac{1}{R_2} + \frac{(n-1)d}{nR_1R_2}\right] \tag{3.1}$$

where n is the refractive index of the lens material; R_1 is the radius of the lens surface that faces closer to the light source; R_2 is the radius of the lens surface farther from the light source; and d is the thickness of the lens.

For paraxial rays, which are rays close to the optical axis and experience minimal deviation when passing through a lens, the relationship between the object distance, image distance, and the lens itself is described by the (Gaussian) thin lens formula [1]:

$$\frac{1}{f} = \frac{1}{S_1} + \frac{1}{S_2} \tag{3.2}$$

where S_1 is the distance from the object to the spherical thin lens, and S_2 is the distance from the lens to the point where the image is formed.

As shown in Figure 3.4, the lens equation can also be expressed in the "Newtonian" form [1]:

$$f^2 = x_1x_2 \tag{3.3}$$

where $x_1 = S_1 - f$, $x_1 = S_1 - f$, and $x_2 = S_2 - f$. x_1 is considered positive when it is positioned to the left of the front focal point F_1, and x_2 is positive when it is positioned to the right of the rear focal point F_2. x_1 and x_2 are either positive or negative. Thin lens imaging properties are shown in Table 3.1.

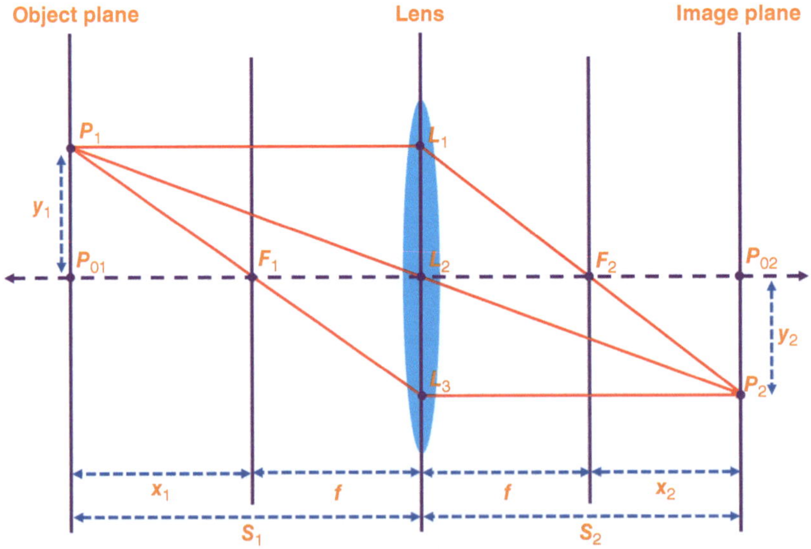

Figure 3.4 Single thin lens imaging. *Source*: Goodphy / Wikimedia Commons / CC BY-SA 4.0 / https://commons.wikimedia.org/wiki/File:Single_Lens_Imaging,_2024-05-30.png / last accessed by October 03, 2025.

Table 3.1 Table for thin lens imaging properties [1].

Lens type	Object location	Image type	Image location	Lateral image orientation	Image magnification	Remark
Converging lens (or positive lens)	$\infty > S_1 > 2f$	Real (rays converging to each image point)	$f < S_2 < 2f$	Inverted (opposite to the object orientation)	Diminished	
Converging lens	$S_1 = 2f$	Real	$S_2 = 2f$	Inverted	Same size	
Converging lens	$f < S_1 < 2f$	Real	$\infty > S_2 > 2f$	Inverted	Magnified	
Converging lens	$S_1 = f$		$\pm\infty$			
Converging lens	$S_1 < f$	Virtual (rays apparently diverging from each image point)	$\lvert S_2 \rvert > S_1$	Erect (same to the object orientation)	Magnified	As an object moves to the lens, the virtual image also gets closer to the lens while the image size is reduced.
Diverging lens (or negative lens)	Anywhere	Virtual	$\lvert S_2 \rvert < \lvert f \rvert$, $S_1 > \lvert S_2 \rvert$	Erect	Diminished	

Source: Goodphy / Wikimedia Commons / CC BY-SA 4.0 / https://en.wikipedia.org/wiki/Lens /last accessed by October 03, 2025.

3.1.2 Optical Aberration

Optical aberration refers to the flaws and deviations from ideal imaging behavior found in optical systems, including lenses, mirrors, telescopes, and other devices that manipulate light. These imperfections arise when light waves passing through an optical element fail to reach a common focal point, a problem typically caused by the design limitations of the system or the intrinsic properties of the materials used. As light transits through these components, various issues can interfere with the precise convergence needed for a sharp and accurate image.

The presence of optical aberrations can manifest in several unwanted ways, leading to distorted or blurred images that fall short of perfection. Such aberrations affect not only the sharpness of an image but also their overall quality in terms of color fidelity and contrast, thus diminishing the accuracy of the visual representation. Whether it be the subtle fringing of colors around an object's edge or an entirely warped image shape, these defects hinder an optical system's ability to faithfully reproduce the scene it captures.

Addressing these aberrations is of paramount importance in various fields where precision and accuracy are essential. Whether in astronomy, where clear celestial observations are critical.

There are several types of optical aberrations, each affecting the image quality in different ways.

3.1.2.1 Defocus

Defocus is a prevalent issue in the field of optics, affecting the clarity and detail of images captured or observed through various optical devices. Essentially, defocus is an optical aberration that results from a misalignment in the focusing process, where the convergence point of light rays does not align with the detection surface, such as the imaging sensor in cameras or the retina of the eye in biological systems. This misalignment leads to a shift in the focus along the optical axis, causing the image to appear blurry or soft rather than sharp and well-defined (Figure 3.5).

In practical terms, defocus manifests as a reduction in image sharpness and contrast. Elements within the scene that should have clear, high-contrast edges take on a softer appearance with gradual transitions between light and dark. This softening effect can obscure fine details, rendering them indistinct or invisible and significantly affecting the overall quality of the image.

To counteract the effects of defocus and restore image clarity, most optical devices are equipped with focus adjustment mechanisms. These adjustments can be manual, requiring the user to make fine tweaks to the focus ring or dial, or they can be automatic, utilizing sensors and motors to adjust the focus rapidly and accurately. In cameras, for instance, autofocus systems are designed to quickly and precisely move the lens elements to achieve the clearest possible image.

Focus adjustment mechanisms are not limited to cameras alone; they are integral to the function of a wide array of optical devices including video cameras, microscopes, telescopes, and binoculars. In all these instruments, achieving an optimal focus is critical to ensure that the observed or captured image is as sharp and clear as

Figure 3.5 A photo of colorful lights with defocus aberration. *Source*: JWCreations / Wikimedia Commons / CC BY-SA 3.0.

possible, enabling the viewer to discern fine details and make accurate observations or recordings.

Beyond mechanical adjustments, some modern optical systems incorporate digital enhancements such as software-driven focus stacking or advanced algorithms to correct defocus. These technologies analyze multiple images taken at different focal distances and combine them to produce a single, sharply focused image. This is particularly useful in fields like macro photography and microscopy, where the depth of field is limited, and achieving a comprehensive focus across the entire scene can be challenging.

3.1.2.2 Spherical Aberration

Spherical aberration is a common optical issue that arises in systems utilizing components with spherical surfaces, and it significantly affects the performance of such systems by degrading image quality. This aberration is especially prevalent in devices with lenses and mirrors that are manufactured with spherical shapes due to the relative ease and cost-effectiveness of producing these surfaces. However, the inherent geometry of spherical surfaces leads to specific optical challenges.

At the core of spherical aberration is the fact that light rays entering or reflecting off a spherical lens or mirror near its periphery do not converge at the same focal point as those rays that pass through or reflect closer to the optical axis or center. As shown in Figure 3.6, this inconsistency in the converging points of light rays results from the geometry of the spherical surfaces, causing the peripheral rays to focus at different points compared to the central rays. Consequently, light rays that strike further from the center are refracted or reflected more strongly, leading to a focus point that is not uniform across the entire surface [2].

The primary result of spherical aberration is a blurred or distorted image. Unlike an ideally focused image where all rays converge at one sharp focal plane, an image impacted by spherical aberration appears with soft edges and lacks crispness and definition. This is particularly problematic in applications requiring precise focusing,

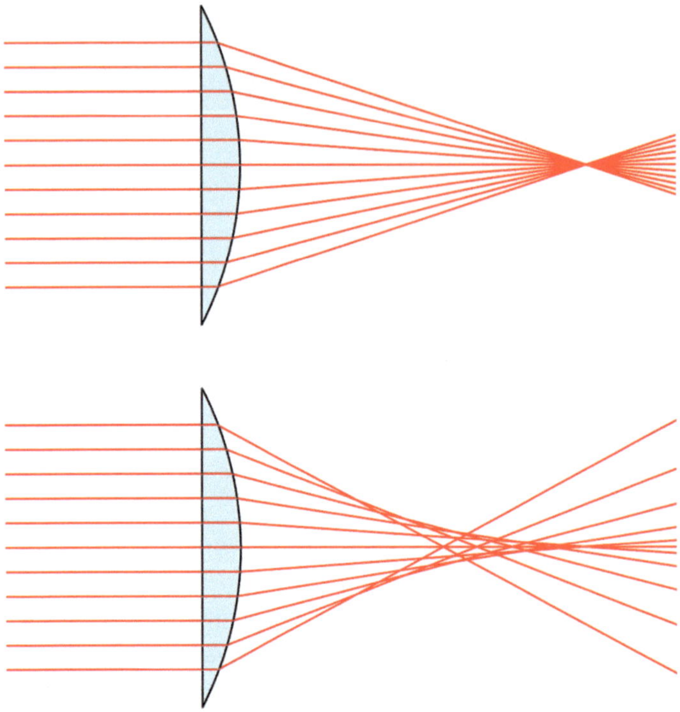

Figure 3.6 At the top is an illustration of an ideal lens free from spherical aberration. The lower example illustrates a practical lens with spherical surfaces. *Source*: Mglg / Wikimedia Commons / CC BY-SA 4.0 / https://commons.wikimedia.org/wiki/File: Spherical_aberration_2.svg / last accessed by October 03, 2025.

such as high-resolution photography, microscopy, and telescope observations, where accuracy in image reproduction is critical.

To combat the detrimental effects of spherical aberration, several corrective strategies have been developed. One common approach is the use of aspherical lens elements. René Descartes demonstrated that lenses with surfaces shaped as specific Cartesian ovals (revolved around the central symmetry axis) can perfectly focus light from a point on the axis or from infinity along the axis direction. This design results in completely aberration-free focusing of light from distant sources. By utilizing carefully chosen aspheric surfaces, optical systems can achieve improved image quality, sharper focus, and reduced aberrations, including spherical aberration. Aspheric lenses are commonly used in various optical instruments and imaging systems to enhance optical performance and minimize aberrations for better image quality.

Another technique to address spherical aberration is the use of aperture stops. By limiting the passage of light to only the central portion of a lens or mirror, an aperture stop can effectively minimize the impact of aberration by blocking the problematic peripheral rays. While this does improve the image quality, it often comes at the cost of reduced light collection, which can be suboptimal in low-light conditions.

In modern optical design, complex combinations of lens elements and advanced coatings are often employed to refine image sharpness and correct various aberrations, including spherical aberration. Additionally, computational methods and post-processing techniques can further enhance image quality by compensating for residual aberrations that physical components might not fully eliminate.

3.1.2.3 Coma

Coma, also known as comatic aberration, is a common optical phenomenon that can significantly impact the performance and image quality of telescopes and other optical systems. This aberration is particularly noticeable when observing point sources of light, such as distant stars, which appear distorted and elongated, resembling comets with characteristic tails. This distortion primarily occurs for off-axis objects, meaning those that are not directly centered within the optical path, and is a result of certain design choices or imperfections in the optical components.

The root cause of coma lies in the variation of magnification across the aperture of an optical system. As shown in Figure 3.7, when light rays from an off-axis point source strike a mirror or lens, they do so at varying angles. For telescopes that incorporate parabolic mirrors – a common design due to their ability to focus parallel rays of light to a single point on the optical axis – the same precision does not extend to off-axis rays. These off-axis rays are not adequately brought to a sharp focus; instead, they form wedge-shaped or comet-like images as they spread out unequally over a focal plane.

In optical systems that span a wide spectral range, such as those involving refractive or diffractive optics, coma can be exacerbated by chromatic aberration,

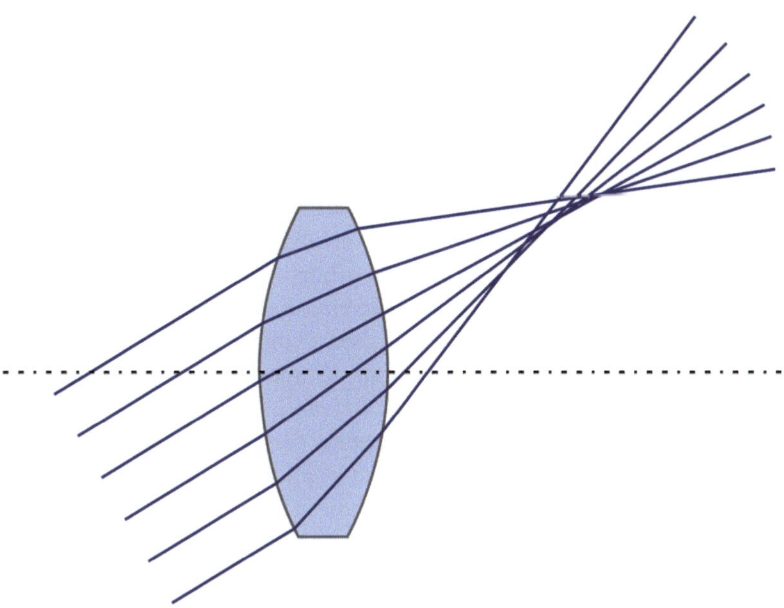

Figure 3.7 Coma of a single lens. *Source*: Wikimedia Commons / CC BY-SA 4.0 / https://commons.wikimedia.org/wiki/File:Lens_coma.svg / last accessed by October 03, 2025.

displaying a dependence on the wavelength of light. This wavelength dependence occurs because light of different wavelengths is bent by slightly different amounts as it passes through lenses, contributing further to the blurring and distortion observed in the final image.

To address the challenges posed by coma, several sophisticated optical design solutions have been developed. For instance, Schmidt and Maksutov telescopes incorporate corrector plates that are designed to adjust the path of light before it reaches the primary mirror, thereby minimizing coma while preserving the desired focus. Similarly, designs like Advanced Coma-Free (ACF) and Ritchey–Chrétien telescopes utilize hyperbolic primary and secondary mirrors to reduce coma without introducing additional spherical aberrations, offering a wider field of view with reduced distortion.

In the context of Newtonian reflectors, which are particularly susceptible to coma due to their straightforward mirror design, coma correctors are often employed. These devices typically use a pair of lenses strategically placed near the focal plane to counter the divergent path of off-axis rays, effectively compressing them back to a coherent point and significantly improving the overall image quality.

For single lenses or simpler lens systems, the minimization of coma is achieved through the meticulous selection of lens surface curvatures, tailored to the specific application and intended wavelength of light. Lenses that effectively eliminate both spherical aberration and coma at a particular wavelength are termed bestform or aplanatic lenses. These lenses rely on carefully calculated curvatures that ensure off-axis rays are focused as accurately as on-axis rays.

By employing these advanced designs and corrective measures, the effects of coma can be managed to great effect, improving image integrity in telescopic observations and other optical applications. This meticulous attention to detail and innovative engineering ensure that we can capture sharp, clear, and accurate images of the night sky and other distant objects with minimal distortion.

3.1.2.4 Astigmatism

In optical systems, astigmatism is a critical aberration that affects the way light rays converge to form images, and addressing it is vital for achieving high-quality visual output. Astigmatism causes rays traveling in two orthogonal planes to converge at different points along the optical axis, leading to blurriness and distortion. This issue is particularly evident when an optical system is used to image a cross: the vertical and horizontal lines will focus sharply at different distances, highlighting one set while blurring the other (Figure 3.8).

The phenomenon of astigmatism in optical systems can be broadly categorized into two main types, each with distinct origins and implications for image quality. The first type, known as third-order or monochromatic astigmatism, occurs primarily for objects or parts of objects located away from the optical axis. It arises even in perfectly symmetrical optical systems and is called "monochromatic" because it is typically analyzed with single-wavelength light. However, the degree of astigmatism can vary significantly with wavelength, which complicates the notion of it being purely a monochromatic issue. This form of astigmatism is a geometric property

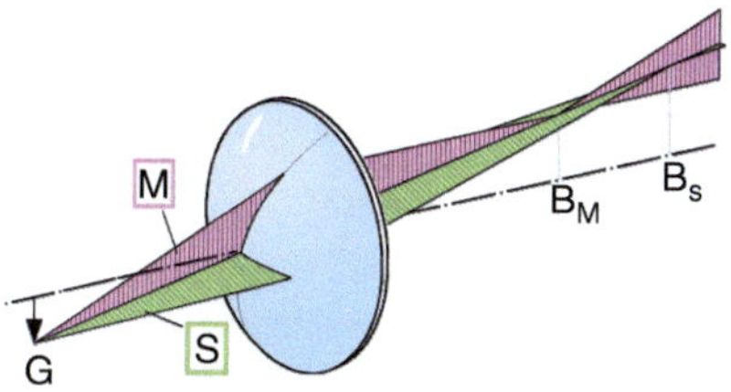

Figure 3.8 Astigmatism causes the image of a single point, such as point G, to focus at different locations depending on the plane within which the light rays travel. For rays in plane M, the image forms at point B_M, whereas for rays in plane S, it converges at point B_S. *Source*: Michael Schmid / Wikimedia Commons / CC BY-SA 4.0 / https://commons.wikimedia.org/wiki/File:MeridionalSagittalPlane.png / last accessed by October 03, 2025.

inherent in dealing with spherical surfaces or other symmetric lens elements and presents challenges in minimizing aberrations in high-performance optical systems, such as telescopes and microscopes.

The second type of astigmatism is due to asymmetry in the optical system. This asymmetry can be intentional, such as when using a cylindrical lens that focuses light differently in one plane compared to the perpendicular plane, or unintentional, arising from manufacturing errors, improper alignment, or installation inaccuracies. Even perfect on-axis points suffer from astigmatism in such systems. This form is particularly relevant in the field of vision science, as the human eye frequently exhibits this type of astigmatism due to irregular shapes of the cornea or lens. Such imperfections can lead to suboptimal focus on the retina, causing blurred vision that necessitates corrective measures like eyeglasses or contact lenses.

Addressing astigmatism in optical systems is crucial for ensuring the desired image quality and visual acuity. Solutions include a careful optical design, high-precision fabrication, and meticulous alignment during assembly. Advanced optical designs might use aspheric lenses to reduce astigmatism and other aberrations. In vision correction, specialized lenses are crafted to counteract the optical errors introduced by astigmatism, thereby restoring clear vision.

As technology progresses, new materials and techniques continue to improve our ability to manage astigmatism. Innovations in adaptive optics, for example, allow the real-time compensation of aberrations, enhancing the performance of telescopes and other advanced optical instruments. Similarly, advancements in manufacturing processes enable more precise control over lens shapes and alignment, leading to better image quality and reduced aberrations.

Ultimately, addressing both forms of astigmatism is essential for ensuring that optical systems, from photographic lenses to telescopes, microscopes, and corrective eyewear, deliver clear, sharp images. This demands a comprehensive approach involving optical design, precision engineering, and rigorous quality control to enhance the system's performance and satisfy user needs effectively. As optical technologies continue to evolve, the quest to minimize astigmatism and other aberrations will drive further innovations and improvements in visual technology.

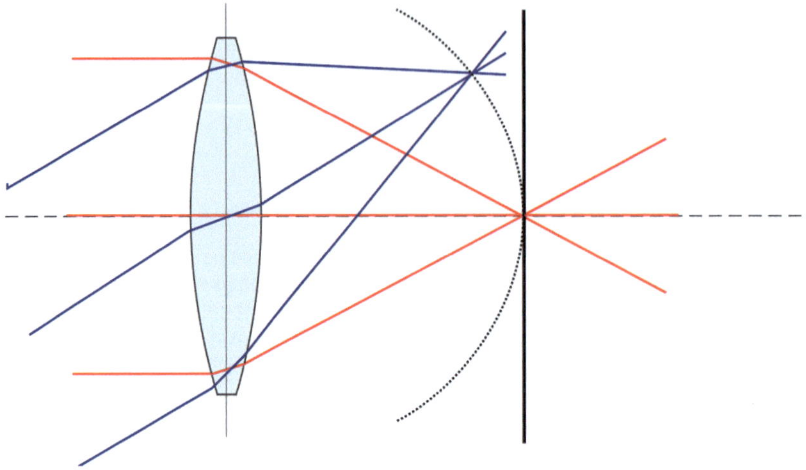

Figure 3.9 Field curvature: the image plane deviates from a flat surface. *Source*: BenFrantzDale / Wikimedia Commons / CC BY-SA 4.0 / https://commons.wikimedia.org/wiki/File:Field_curvature.svg / last accessed by October 03, 2025.

3.1.2.5 Field Curvature

Field curvature is an optical aberration that occurs in imaging systems when the sharpest focus of the image forms on a curved surface, known as the Petzval surface [3], rather than on a flat plane like a camera sensor or photographic film (Figure 3.9). This imperfect focusing leads to a challenge: if one area of the image field, such as the center, is in perfect focus, other areas, like the edges or corners, may appear blurred, and vice versa [4]. This issue is especially prevalent in optical systems designed with wide fields of view, such as wide-angle lenses or telescopes, where maintaining focus uniformity across the frame is challenging.

The effect of field curvature is that it causes portions of the image to look out of focus, depending on where the focal plane intersects the image field. This can result in a visual effect where the central portion of a photo or an image from a telescope appears sharp while the periphery becomes softer, blurring details that are not in the central field of view. Conversely, focusing on the edges can blur the center, which is problematic for applications requiring high-resolution, consistent sharpness across the entire image, such as in landscape photography or when capturing large prints.

Addressing field curvature often involves considering complex optical designs. Lens manufacturers might incorporate additional lens elements, such as aspherical lenses or field flatteners, to counteract the natural curvature of the image field. These components work by adjusting the light paths so that they converge onto a single, flat plane, thus ensuring the entire image remains in sharp focus.

In advanced optical systems, such as those used in professional photography, cinematography, or scientific analysis, mitigating field curvature is crucial for achieving the precise image quality demanded by these disciplines. Furthermore, understanding field curvature is essential in the design of optical systems used in astronomy, microscopy, and any other fields relying on detailed, wide-field imaging. By managing field curvature effectively, optical engineers ensure that their systems deliver uniformly sharp images, hence meeting the various demands of visual accuracy and detail across different sectors.

3.1.2.6 Image Distortion

Image distortion is an optical aberration that results in the alteration of an image's shape compared to the original object, occurring when lenses fail to maintain consistent magnification across the field of view. This inconsistency causes parts of the image to stretch or compress disproportionately, producing various forms of distortion. This phenomenon is distinct from other optical aberrations, such as spherical aberration, coma, chromatic aberration, field curvature, and astigmatism. While those aberrations primarily affect the sharpness and clarity of an image without altering the geometry of objects, distortion directly impacts the object's shape, leading to visual misrepresentation.

While distortion can manifest in various irregular patterns, the most common types are radially symmetric, or approximately so, due to the inherent symmetry of photographic lenses. These distortions are typically categorized as barrel distortion, pincushion distortion, and mustache distortion (Figure 3.10).

Barrel distortion: This type of distortion occurs when the magnification decreases with distance from the optical axis, causing straight lines to bow outward, similar to the shape of a barrel. This is commonly seen in wide-angle lenses and causes the edges of an image to appear "stretched." Barrel distortion is often encountered in landscape photography and can give images a bubbled appearance.

Pincushion distortion: In contrast, pincushion distortion happens when magnification increases with distance from the optical axis. This effect causes straight lines to bend inward toward the center, making the image edges appear pinched, resembling the sides of a pincushion. Pincushion distortion is more common in telephoto lenses and can lead to images that have a squeezed-in look, particularly affecting portraiture and telephoto shots.

Mustache distortion (complex distortion): This type of distortion involves elements of both barrel and pincushion distortion, which can occur in more sophisticated

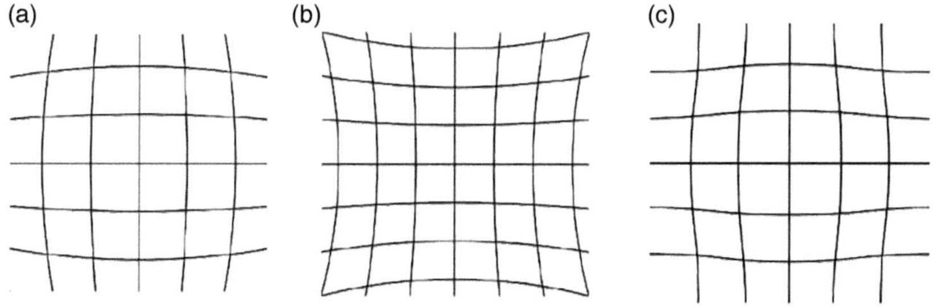

Figure 3.10 Examples of radial distortions. (a) Barrel. (b) Pincushion. (c) Mustache. *Source*: WolfWings / Wikimedia Commons / CC BY-SA 4.0 / https://en.wikipedia.org/wiki/Distortion_(optics)#/media/File:Barrel_distortion.svg / last accessed by October 03.

lens systems or at varying focal lengths in zoom lenses. This can lead to an image that has both rounded and pinched characteristics simultaneously.

Correcting distortion is important in applications where accurate image representation is critical. Techniques to address distortion include using specially designed lenses, applying in-camera adjustments, and leveraging post-processing software. By managing distortion, photographers and videographers can ensure that images maintain their geometric fidelity, providing accurate and visually appealing results.

3.1.2.7 Chromatic or Color Aberration

Chromatic aberration, also referred to as chromatic distortion or color aberration, is an optical issue that occurs when a lens fails to focus all colors of light to the same focal point. This phenomenon is primarily caused by dispersion – the dependence of a lens material's refractive index on the wavelength of light. As each color of light has a different wavelength, they are refracted, or bent, by varying amounts as they pass through a lens (Figure 3.11).

In practice, this means that light rays of different colors do not converge at the same location, leading to an image that lacks sharpness and color accuracy. The disparity in refraction causes colors to focus at different depths or with varying magnifications, resulting in visible color fringing or shifts in an image. This is often evident as the fringes of unnatural colors like purple, blue, or green along the boundaries of objects in photographs, particularly in high-contrast areas where the light and dark regions meet.

The refractive index of most transparent materials decreases as the wavelength of light increases, affecting how light is focused. Consequently, different colors are focused at different points in relation to the lens, creating aberrations that can impact image quality.

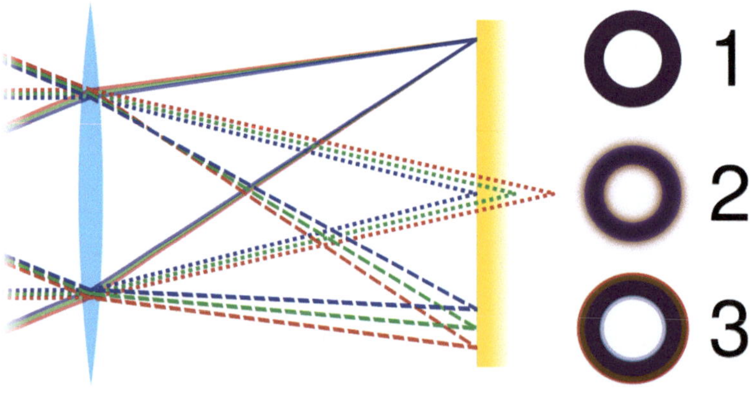

Figure 3.11 Comparison of an ideal image of a ring (1) and ones with only axial (2) and only transverse (3) chromatic aberration. *Source*: Cmglee / Wikimedia Commons / CC BY-SA 4.0 / https://commons.wikimedia.org/wiki/File:Comparison_axial_lateral_chromatic_aberration.svg / last accessed by October 03, 2025.

Chromatic aberration can manifest in two main ways. Longitudinal chromatic aberration, also known as axial chromatic aberration, occurs when different wavelengths of light, each representing distinct colors, are focused at varying distances along the optical axis of the lens. This results in a misalignment where colors such as red, blue, and green do not converge simultaneously, creating color fringing that appears as the glowing halos of color around objects, especially noticeable in areas with stark contrast. For instance, white subjects against a dark background may exhibit blurred edges tinged with hues that appear like spectral outlines, thus reducing image's sharpness and overall color fidelity.

On the other hand, lateral chromatic aberration, or transverse chromatic aberration, takes place when light rays of different colors are refracted but do not align properly across the image plane. Instead, the different colors are focused at various lateral positions in the image, principally affecting the periphery. This misalignment leads to color distortion predominantly at the image edges, creating color fringing that does not blur like longitudinal aberration but rather shifts the colors side-by-side, resulting in an image where the outer portions exhibit unnatural color separation [5].

Understanding and addressing chromatic aberration is essential for achieving high-quality images in photography and optical systems. Techniques such as lens design improvements, optical coatings, and digital correction methods can help minimize or correct chromatic aberration effects in images.

To mitigate chromatic aberration, lens designers employ advanced optical techniques aimed at aligning the focal points of different colors, thereby improving image clarity and color accuracy. One of the primary methods to achieve this involves the design and use of achromatic and apochromatic lenses.

Achromatic lenses are specifically crafted to reduce chromatic aberration by combining two or more lens elements made from different types of glass with varying dispersion properties. Typically, an achromatic lens consists of a converging (positive) lens made of crown glass and a diverging (negative) lens made of flint glass (Figure 3.12). Each glass type has distinct refractive indices – the degree to which they bend light – and when paired, they can correct for the different focal points of blue and red light, bringing them closer together. The merging of these two elements effectively reduces the overall color fringing seen in images. While an achromatic lens primarily addresses two wavelengths (usually blue and red), creating what is known as a "doublet," it significantly minimizes the longitudinal chromatic aberration that can affect simple lenses. The result is an optical component that provides clearer and more color-accurate images across a portion of the visible spectrum [6].

Apochromatic lenses take the concept a step further. These lenses are engineered to bring three wavelengths of light (commonly red, green, and blue) into a common focus. This is achieved by using more than two glass elements and carefully optimizing each for dispersion across a broader range of wavelengths. Apochromats use advanced materials and sophisticated designs, often incorporating additional lens elements, to tackle both longitudinal and lateral chromatic aberrations comprehensively. The design complexity of apochromatic lenses results in the

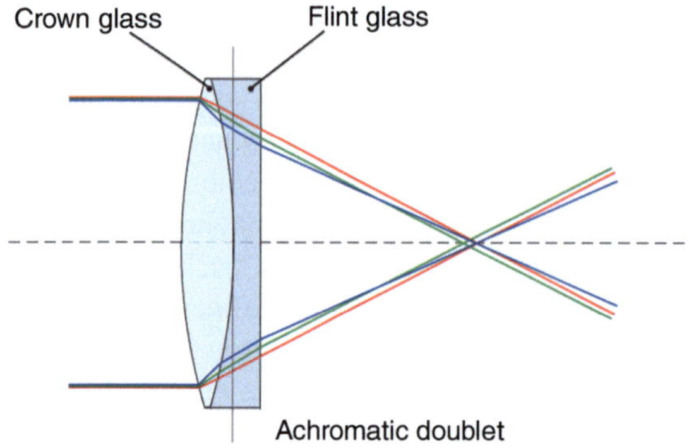

Figure 3.12 Schematic diagram of an achromatic doublet. *Source*: DrBob / Wikimedia Commons / CC BY-SA 4.0 / https://commons.wikimedia.org/wiki/File:Lens6b-en.svg / last accessed by October 03, 2025.

superior correction of chromatic aberration over a wider spectrum compared to achromatic lenses. This makes them particularly valuable in applications requiring high precision and resolution, such as microscopy and high-end photography, where color fidelity and sharpness are crucial.

3.2 Light Source and Characteristics

In OET systems, the choice of light source is a critical component that significantly influences the system's performance and versatility. Unlike other optical trapping techniques, OET does not require coherent light sources such as lasers. This is because OET systems primarily rely on the creation of electric fields to manipulate particles, rather than using radiation pressure from a focused beam. As a result, the requirements for the light source are less stringent in terms of coherence and intensity.

3.2.1 LCD

Liquid-crystal displays (LCDs) are a fascinating intersection of optical science and technology, enabling a wide array of modern electronic devices to present visual information efficiently. These flat-panel displays or optical devices leverage the unique light-modulating properties of liquid crystals alongside polarizers to display information vividly and clearly.

Liquid crystals themselves do not emit light. Instead, they depend on a secondary light source, such as a backlight, to produce visible images. This reliance on external lighting is integral to LCD functionality, as it determines how images, whether in color or monochrome, are portrayed on the screen. This requirement distinguishes LCDs from self-emissive display technologies like organic light-emitting diodes (OLEDs) [7].

LCDs can render arbitrary images, such as those seen on a computer or smartphone screen, by utilizing a matrix of small, discrete pixels. Alternatively, they can display fixed images with a low information content, like digits on a digital clock or simple preset words. Despite the diversity in applications, both types of displays utilize the same core technology. For arbitrary images, the intricate grid of pixels allows for high-resolution and detailed graphics, while simpler fixed-image displays maintain larger elements to show basic information clearly and reliably.

The construction of an individual LCD pixel is sophisticated. Each pixel typically consists of a thin layer of liquid crystal molecules sandwiched between two transparent electrodes, which are often made from indium tin oxide (ITO) due to its excellent conductivity and transparency. These electrodes are paired with two polarizing filters assembled with their transmission axes perpendicularly aligned. In their natural state, and without an applied electric field, the alignment of the liquid crystal molecules is dictated by the alignment layers on the electrode surfaces.

As shown in Figure 3.13, a common structure employed in LCDs is the twisted nematic (TN) configuration. In a TN display, the alignment directions on the two electrode surfaces are set perpendicularly, causing the liquid crystal molecules to adopt a spiral or helix formation. This twisting of molecules affects the polarization of incoming light, allowing it to rotate as it passes through, giving the pixel a distinct gray appearance when viewed. However, when a voltage is applied across the liquid crystal layer, the electric field influences the molecular alignment, essentially "untwisting" the molecules. This results in little to no change in the polarization of light as it traverses the layer, rendering the pixel dark as the perpendicular polarizing filter blocks the light completely [8].

By manipulating the voltage applied to each pixel, an LCD can control the amount of light that eventually reaches the viewer's eyes. This variable control over light

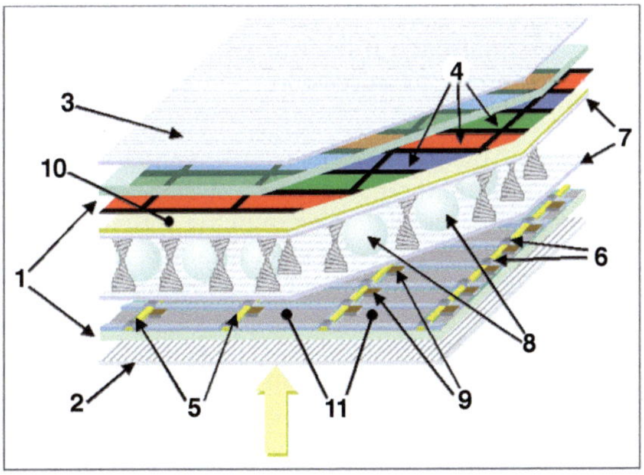

Figure 3.13 Layers of a color TFT LCD. 1: Glass plates. 2 and 3: Horizontal and vertical polarizers. 4: RGB color mask. 5 and 6: Horizontal and vertical command lines. 7: Polymer layer. 8: Spacers. 9: Thin film transistors. 10: Front electrode. 11: Rear electrode. *Source*: Lozère / Wikimedia Commons / CC BY-SA 4.0 / https://commons.wikimedia.org/wiki/File:Color_TFT-LCD_Layout.png / last accessed by October 03, 2025.

passage allows the creation of different shades of gray, and in conjunction with color filters, full-color images.

Beyond the fundamental technology of modulating light, LCDs require external light to generate visible images, as they do not produce light themselves. In most modern LCDs, particularly those employing the active-matrix technology, this light source is a backlight mounted behind the glass layers of the display. Such backlit LCDs are common in devices where high brightness and clarity are essential, such as in televisions, monitors, and portable electronics.

Contrastingly, some LCDs employ a reflective mode, using a mirror-like surface or reflective film at the rear to capitalize on ambient light. These passive LCDs are often found in lower-power devices like digital watches. Transflective LCDs offer a versatile combination of both approaches, operating efficiently in varying lighting conditions by incorporating the characteristics of both backlit transmissive displays and reflective displays.

This ability to adapt to different lighting scenarios while maintaining energy efficiency is a part of what makes the LCD technology so versatile and pervasive in electronic displays today, supporting a wide spectrum of applications from simple digital readouts to complex graphical interfaces.

3.2.2 DMD

A digital micromirror device (DMD), as illustrated in Figure 3.14, is a crucial component in digital projectors. It comprises an array of millions of tiny, individually tiltable mirrors that form the pixels in the projected images. Each DMD chip contains hundreds of thousands of these minute mirrors, arranged in a precise rectangular pattern that corresponds to the pixels of an image. Fabricated using advanced surface micromachining techniques, these mirrors are made of aluminum and measure approximately 16 μm in diameter. They are mounted on yokes and connected to fixed support posts through compliant torsion hinges, allowing each mirror to tilt by ±10–12 degrees. The tilt mechanism enables each mirror to switch between an

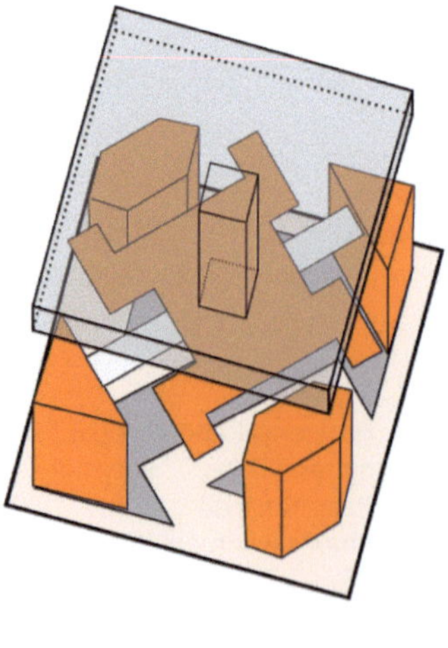

Figure 3.14 Diagram of a digital micromirror illustrating the mirror mounted on a suspended yoke, with the torsion spring running from the bottom left to the top right (light grey). The electrostatic pads of the memory cells are positioned below, at the top left and bottom right. *Source*: Egmason / Wikimedia Commons / CC BY-SA 4.0 / https://commons.wikimedia.org/wiki/File:Digital_micromirror2.svg / last accessed by October 03, 2025.

"on" and an "off" position, creating bright or dark pixels, respectively, by reflecting light from the projector bulb toward or away from the lens. Greyscales are achieved through the binary pulse-width modulation, a process where mirrors switch on and off rapidly. This rapid toggling adjusts the time light reflects from the mirror to the screen, producing various shades of gray. Modern DMD chips can offer up to 1024 shades of gray, equivalent to a 10-bit depth [9].

Each mirror's position is controlled by two pairs of electrodes inducing electrostatic attraction, with one pair acting on the yoke and another directly on the mirror. Typically, equal bias charges are applied to both sides of the mirror to maintain its position, as the stronger attraction force from the side closer to the electrodes holds the mirror steady [10].

The operation of these mirrors is intricately linked to static random-access memory (SRAM) cells situated beneath each pixel. These cells store the desired state for each mirror and connect to the controlling electrodes. Once all SRAM cells are loaded with data, the bias voltage is temporarily removed, allowing the stored charges to adjust the mirror position accordingly. Reinstituting the bias voltage locks the mirrors in place, with the entire array adjusting for the next display cycle simultaneously. This synchronized approach minimizes the voltage requirements from the SRAM cells and enhances the precision of mirror movements, contributing to the cinematic quality of motion pictures by ensuring smooth, accurate transitions during projection.

In addition to typical visual applications, DMDs serve critical roles in other fields such as 3D scanning, where the precise control over light reflection is necessary, and in advanced manufacturing processes like lithography. Their versatility, resilience, and precision engineering make DMDs a pivotal element in the development and evolution of modern optical systems.

3.2.3 LCoS

LCoS, which stands for liquid crystal on silicon, is a type of reflective display technology commonly used in projectors, head-mounted displays, and some high-resolution televisions. Unlike typical transmissive LCDs that allow light to pass through the panel, LCoS utilizes a reflective setup with liquid crystal material applied directly onto a silicon backing. This arrangement facilitates precise light modulation for each pixel, as it relies on reflection rather than direct transmission through the display.

In terms of structure (Figure 3.15), an LCoS system typically comprises a light source, such as a high-intensity lamp or an LED array, along with a beam-splitting setup and the LCoS panel itself. The panel contains a silicon backplane coated with a liquid crystal layer situated between two glass substrates. When light from the source is directed onto the LCoS panel, the liquid crystals modulate the light according to the incoming image data. Each liquid crystal cell acts as a pixel, adjusting its orientation to control how light is reflected back [11].

This reflected light, once modulated by the liquid crystals, is then directed through optical components that merge the red, green, and blue light to produce a full-color image. The meticulous control over each pixel enables LCoS to deliver

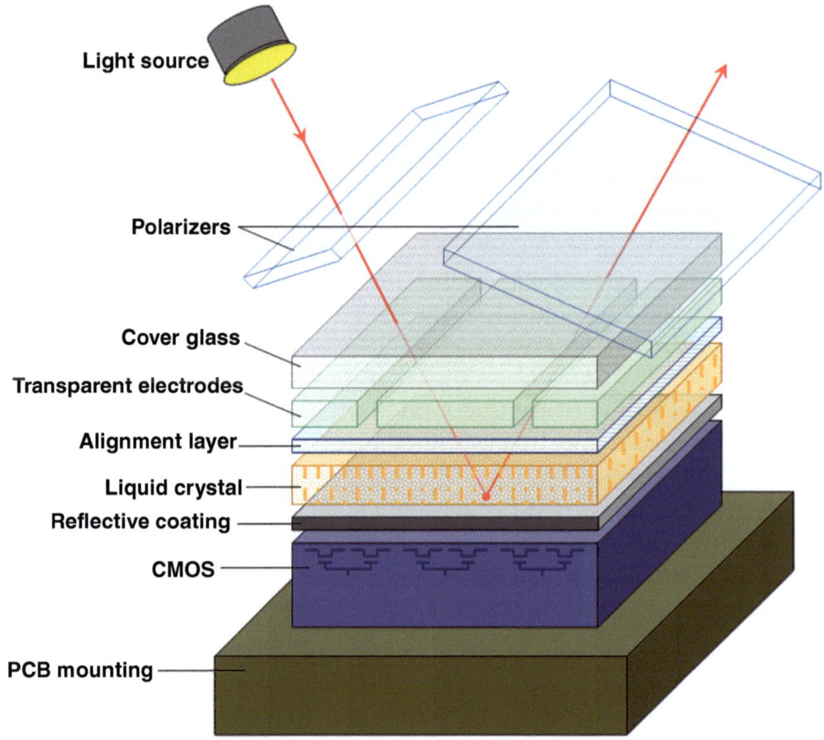

Figure 3.15 LCoS microdevice scheme. *Source*: Aesopus / Wikimedia Commons / CC BY-SA 4.0 / https://upload.wikimedia.org/wikipedia/commons/4/4e/LCOS-eng.png/ last accessed by October 03, 2025.

high-resolution and exceptionally sharp images, characterized by excellent contrast and color fidelity due to the reflective nature of the panel.

The LCoS technology is particularly valuable for its ability to generate high-resolution images because it supports a dense pixel arrangement due to the compact size of the liquid crystal cells [12]. This leads to images that are smooth and detailed, without the screen-door effect that can sometimes be seen in other display technologies.

The advanced image quality of LCoS is utilized in a variety of applications, including high-definition home theater projectors, professional video projectors, head-mounted displays for virtual or augmented reality, and certain high-resolution television models. Its precision and detail make it highly suitable for fields that demand impeccable visual quality, such as scientific visualization, medical imaging, and digital cinema. Overall, LCoS stands out as a powerful display technology capable of delivering superior image fidelity for both consumer and professional uses.

3.2.4 LED

An LED (Figure 3.16) is a remarkable semiconductor device capable of generating light through its unique operation as electric current flows through it. At the core of

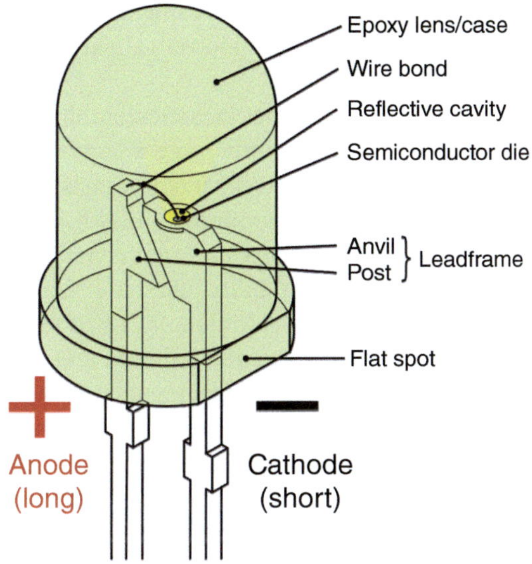

Figure 3.16 Parts of a conventional LED. *Source*: Inductiveload / Wikimedia Commons / CC BY-SA 4.0 / https://commons.wikimedia.org/wiki/File:LED,_5mm,_green_(en).svg / last accessed by October 03, 2025.

an LED's operation is the interaction within its semiconductor material, where electrons recombine with electron holes. This recombination process releases energy in the form of photons, which are particles of light. The particular wavelength, and therefore color, of the emitted photons is characteristic of the energy required for electrons to cross the band gap of the semiconductor material. This allows LEDs to produce a wide spectrum of colors, determined by the specific materials and structure of the semiconductor.

To produce white light, which is not a single color but a combination, different strategies are employed. One common method involves using multiple semiconductor materials, each emitting different colors, which when combined, give the appearance of white light. Alternatively, a single semiconductor can be coated with a light-emitting phosphor layer, which absorbs some of the light and re-emits it at a different wavelength, again resulting in the perception of white light.

Initially, LEDs found widespread use as indicator lights in electronic devices, gradually replacing small incandescent bulbs due to their efficiency and longevity. They were also integral to seven-segment displays used in calculators and digital clocks. As technology evolved, LEDs capable of emitting not only visible light but also ultraviolet (UV) and infrared wavelengths were developed, significantly broadening their application scope. Today's LEDs range from high-output devices used in lighting applications to low and intermediate-output types used in electronics and other devices. White LEDs, in particular, have gained popularity in both indoor and outdoor lighting due to their efficiency and brightness.

The versatility of LEDs extends beyond traditional lighting. Their development has paved the way for new display technologies, including large-scale digital displays and sophisticated sensors. The rapid switching speeds of LEDs make them invaluable in modern communication systems where fast and reliable signal transmission is crucial. As a result, LEDs have become ubiquitous, with applications spanning diverse fields such as aviation lighting, decorative fairy lights, flexible strip lights, automotive headlamps, advertising displays, general-purpose lighting, traffic signals, camera flashes, illuminated wallpaper, horticultural grow lights, and medical devices [13].

Compared to traditional incandescent light sources, LEDs offer a host of advantages. They are significantly more energy-efficient, consuming less power for the same light output. This efficiency translates into lower operational costs and a smaller environmental footprint. LEDs have a longer lifespan, reducing the need for frequent replacements and maintenance. Their robust construction provides enhanced physical durability, allowing them to withstand vibration, shock, and other physical stresses. Additionally, LEDs are compact, lending themselves well to a wide range of form factors in device and system designs. Their rapid switching capability means they can turn on and off nearly instantaneously, which is critical in applications like data communications and lighting effects.

However, LEDs are not without their challenges. They operate most effectively within specific voltage limits and typically require direct current (DC) power, as opposed to alternating current (AC), to function properly. This can necessitate additional power conversion components in systems primarily designed for AC operation. Although advancements have been made, LEDs can still struggle with consistent light output when powered by pulsing DC or AC sources. Moreover, their performance can be affected by low maximum operating and storage temperatures, limiting their use in some extreme environments.

3.2.5 Laser

A laser (Figure 3.17), which stands for "light amplification by stimulated emission of radiation," is a device designed to emit light with remarkable coherence and intensity. The light produced by lasers is noted for being monochromatic, meaning it consists of a single wavelength or color, and highly collimated, allowing the beam to remain narrow over long distances. These characteristics result from the fundamental processes and components that constitute a laser system.

At the heart of a laser's operation is the principle of stimulated emission. When atoms or molecules in the laser's gain medium are excited to higher energy levels, they can return to lower energy states, releasing photons in the process. If a photon of the appropriate energy interacts with an excited atom, it can prompt the emission of another photon. This emitted photon is coherent with the original – it shares the

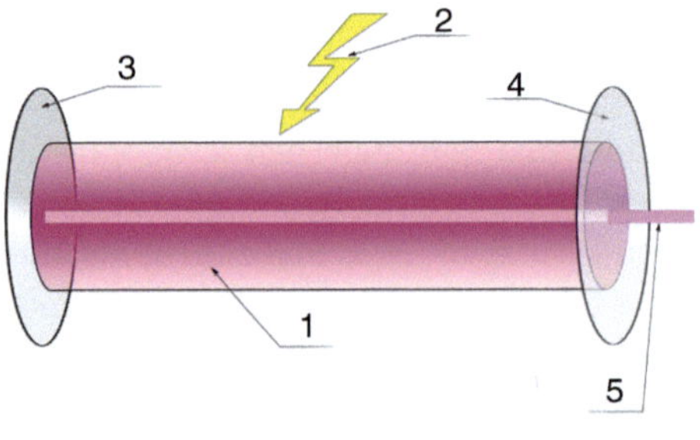

Figure 3.17 Components of a typical laser: 1. Gain medium. 2. Laser pumping energy. 3. High reflector. 4. Output coupler. 5. Laser beam. *Source*: Tatoute / Wikimedia Commons / CC BY-SA 4.0 / https://commons.wikimedia.org/wiki/File:Laser.svg / last accessed by October 03, 2025.

same phase, frequency, and direction. This process effectively amplifies light, as each photon can stimulate further emissions [14].

The laser contains an optical cavity, usually established by placing a gain medium between two mirrors. One mirror is fully reflective, and the other is partially transmissive to allow some light to escape. This cavity efficiently reflects photons, enabling them to pass through the gain medium multiple times, each instance causing more stimulated emissions and thereby increasing the light's intensity [15].

An essential requirement for laser operation is achieving a population inversion, where more particles are in the excited state than in the ground state. This inversion is typically accomplished by pumping energy into the system from an external source like an electrical current, a flashlamp, or even another laser. By maintaining the population inversion, the laser can sustain a continuous stimulated emission.

There are several types of lasers, each defined by its gain medium. Solid-state lasers, for example, use a solid material like neodymium-doped yttrium aluminum garnet (Nd:YAG) or ruby. Gas lasers, such as those involving helium–neon or carbon dioxide, create continuous and powerful beams, suitable for a variety of uses. Semiconductor lasers, or laser diodes, are compact and efficient, making them ideal for consumer electronics, telecommunications, and simple pointers. Excimer lasers use reactive gases like fluorine, producing UV light and playing key roles in microfabrication and eye surgeries. Dye lasers utilize organic dyes dissolved in liquid, offering the unique capability to tune their output wavelengths.

The applications for lasers are vast owing to their precision and adaptability. In communications, lasers play a crucial role in fiber optic networks, transmitting data over long distances with minimal loss. Medical applications are equally impressive; lasers are employed in surgeries, noninvasive procedures for skin conditions, and precise diagnostics. In industry, lasers aid in cutting, welding, and marking materials with unmatched precision. Scientific research benefits tremendously from lasers, which explore atomic behavior, enable the study of chemical reactions, and analyze the properties of new materials. Even in everyday life, lasers have become ubiquitous, being central to the functioning of CD/DVD players and barcode scanners.

Lasers are exceptional because of their ability to deliver focused, high-intensity light. This focus comes from coherence, allowing light waves to align perfectly in space and time. The high degree of monochromaticity contributes to the laser's ability to target specific wavelengths for precise interactions with materials, leading to innovations in practically every technical field. Whether for cutting-edge research, routine industrial processes, or consumer electronics, lasers hold an indispensable position in modern technology.

3.2.6 Mercury-Vapor Lamp

A mercury-vapor lamp (Figure 3.18) is a gas discharge lamp that produces light by creating an electric arc through vaporized mercury. These lamps are known

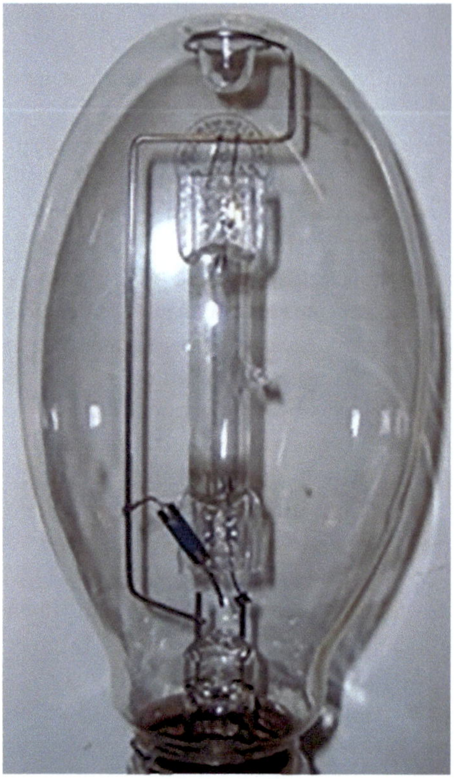

Figure 3.18 A typical high-pressure mercury lamp. *Source*: Tylercat5 / Wikimedia Commons / Public domain.

for emitting a bright and intense light, making them a popular choice in both industrial and commercial lighting applications. The operating mechanism of a mercury-vapor lamp hinges on the interaction between electricity and mercury in a sealed environment.

The construction of a mercury-vapor lamp involves a sealed tube made of glass or quartz, containing a small amount of liquid mercury and an inert gas, commonly argon, at a low pressure. Inside this tube, two main electrodes are placed at opposite ends, connected to an external electrical power source. When the lamp is switched on, a high voltage is applied across these electrodes, which ionizes the argon gas, initiating an electric arc. This arc provides the necessary energy to vaporize the mercury also contained within the tube [16].

As the electric current passes through the tube, the vaporized mercury becomes ionized and conducts electricity more effectively. The electrical discharge in the vaporized mercury produces light that predominantly falls in the UV spectrum, with some visible light as well. To enhance the amount of visible light and to modify the light output for practical uses, many mercury-vapor lamps incorporate a phosphor coating on the inner surface of the outer bulb. This phosphor layer absorbs the UV light and re-emits it as visible light, often adjusting the color balance to create a whiter, more visually appealing light [17].

Mercury-vapor lamps boast high efficiency in terms of lumens per watt, offering substantial light output for the energy consumed, though they are not as energy-efficient as modern lighting technologies like LEDs. One of their advantages is their long operational life, often exceeding several thousand hours, which makes them a cost-effective choice for environments requiring prolonged or continuous illumination.

The light emitted without a phosphor coating has a characteristic bluish hue due to the wavelengths of the visible portion of mercury's emission spectrum. However, with a phosphor coating, the color can be adjusted to a more balanced white light, making these lamps more suitable for a wider range of lighting applications. Mercury-vapor lamps have traditionally been used for outdoor lighting, such as street lamps, floodlights for stadiums, and lighting for large industrial spaces. They can also be found in photographic studios and art installations where specific lighting effects are needed.

However, there are some drawbacks to using mercury-vapor lamps. The inclusion of mercury, a toxic substance, presents environmental and safety concerns, particularly regarding disposal and potential ecological impact. Additionally, these lamps have a startup time and require several minutes to reach their full brightness after being powered on, which can be a limitation in situations where instant light is necessary.

References

1 Hecht, E. (2017). *Optics*, 5the. Pearson.

2 Smith, T.T. (1922). Spherical aberration in thin lenses. *Physical Review* 19 (3): 276.

3 Riedl, M.J. (2001). *Optical Design Fundamentals for Infrared Systems*. SPIE press.

4 Fitzgibbon, A. W., (2001). Simultaneous linear estimation of multiple view geometry and lens distortion. The 2001 IEEE Computer Society Conference on Computer Vision and Pattern Recognition (8–14 December 2001), IEEE.

5 Marimont, D.H. and Wandell, B.A. (1994). Matching color images: the effects of axial chromatic aberration. *Journal of the Optical Society of America A.* 11 (12): 3113–3122.

6 Amako, J., Nagasaka, K., and Kazuhiro, N. (2002). Chromatic-distortion compensation in splitting and focusing of femtosecond pulses by use of a pair of diffractive optical elements. *Optics Letters* 27 (11): 969–971.

7 Kawamoto, H. (2002). The history of liquid-crystal displays. *Proceedings of the IEEE* 90 (4): 460–500.

8 Ibrahim, D. (2012). *Using LEDs, LCDs and GLCDs in Microcontroller Projects.* Wiley.

9 Brennesholtz, M.S. and Edward, H. (2008). *Projection Displays*, 57–69. Wiley.

10 Horenstein, M.N., Pappas, S., Fishov, A., and Bifano, T.G. (2002). Electrostatic micromirrors for subaperturing in an adaptive optics system. *Journal of Electrostatics* 54 (3–4): 321–332.

11 Zhang, Z., You, Z., and Chu, D. (2014). Fundamentals of phase-only liquid crystal on silicon (LCOS) devices. *Light: Science & Applications* 3 (10): e213.

12 Johnson, K.M., McKnight, D.J., and Underwood, I. (1993). Smart spatial light modulators using liquid crystals on silicon. *IEEE Journal of Quantum Electronics* 29 (2): 699–714.

13 Pelaez, E. A., and Villegas, E. R. (2007). LED power reduction trade-offs for ambulatory pulse oximetry, *29th Annual International Conference of the IEEE Engineering in Medicine and Biology Society* (22 October 2007), IEEE, pp. 2296–2299.

14 Bertolotti, M. (2004). *The History of the Laser*. CRC Press.

15 Renk, K.F. (2012). *Basics of Laser Physics*. Springer.

16 Schiler, M. (1997). *Simplified Design of Building Lighting*. Wiley.

17 Perkin, F.M. (1911). Mercury vapour lamps and action of ultra violet rays. *Transactions of the Faraday Society* 6: 199–204.

4

Electro-optical Interactions

The photovoltaic effect serves as the primary mechanism for generating nonuniform electric fields in the OET technology. This chapter offers a detailed explanation of the physical principles underlying the photovoltaic effect and describes how optoelectronic thin films in OET systems absorb photons to generate these nonuniform electric fields. Additionally, the chapter discusses the selection of optoelectronic materials and the fabrication processes involved in creating OET chips. Various materials and fabrication methods can produce optoelectronic thin films with distinct properties and structures tailored to the needs of different applications. Finally, the chapter presents several methods for evaluating the performance of these optoelectronic thin films.

4.1 Interaction Between Light and Electric Fields

4.1.1 Photoelectric Effect

The main mechanism for creating a nonuniform electric field in OET is the photoelectric effect. This effect occurs when light shines on a metal material, causing changes in the electrical properties of the material's surface. While German physicist Hertz first observed this phenomenon in 1887, it was not until 1905 that Einstein, drawing on Planck's quantum theory, proposed the photon hypothesis to successfully explain the photoelectric effect. Einstein's groundbreaking work on this topic earned him the 1921 Nobel Prize in Physics [1].

The photoelectric effect can be classified into photoelectron emission, photoconductive effect, and photovoltaic effect. The first one is also known as the external photoelectric effect, while the latter two are referred to as the internal photoelectric effects. The external photoelectric effect is the phenomenon where electrons, stimulated by light, escape from the surface of a substance and form electrons in vacuum. The emitted electrons are termed as photoelectrons. The internal photoelectric effect is the phenomenon that the charge carriers (free electrons or holes) generated by light excitation still move within the substance, resulting in a change in the conductivity of the substance or the generation of a photovoltaic effect.

Optoelectronic Tweezers for Manipulation at the Micro and Nano Scale, First Edition.
Lin Feng, Menglu Tan, Ao Wang, and Jiaying Zhang.
© 2026 WILEY-VCH GmbH. All rights reserved, including rights for text and data mining and training of artificial intelligence technologies or similar technologies. Published 2026 by WILEY-VCH GmbH.

According to Einstein's photoelectric effect equation, the escape velocity of electrons under the action of an electric field is:

$$h\nu = \frac{1}{2}mv^2 + W \tag{4.1}$$

where h represents Planck's constant, $h = 6.63 \times 10^{-34}$ J/s, ν represents the frequency of the incident photon, $\frac{1}{2}mv^2$ represents the initial kinetic energy of the photoelectron exiting the object, and W represents the escape work, i.e. the minimum energy required for the electron to escape from the metal surface.

4.1.2 Photoconductive Effect

In 1873, British scientist W. Smith pointed out that the resistance of materials will change with the intensity of light [2]. This discovery was one of the first important experimental observations in the field of semiconductors, revealing the sensitivity of semiconductors to light, a phenomenon also known as the photoconductivity effect. The photoconductivity effect is one of the two internal photoelectric effects. When the light is irradiated to the semiconductor material, if the photon energy is greater than the binding energy of the electrons in the material, the photon energy is absorbed by the photoelectric material, and the electrons in the material obtain energy from the valence band (the energy level of the valence electrons) to the conduction band (the energy level of the free electrons), becoming free electrons, causing the increase of the carrier concentration, leaving holes in the valence band, forming an electron–hole pair. Under the influence of the electric field generated by the external electrode, the electron–hole pair transmits in the opposite direction, creating an electric current and reducing the resistivity of the material.

The carrier production rate $G(x)$ of light in a semiconductor along the illumination direction x is defined as the number of electron–hole pairs produced by absorbing light per unit time and per unit volume:

$$G = \alpha N_0 \exp(-\alpha x) \tag{4.2}$$

$$\alpha = \frac{4\pi}{\lambda}\kappa_s \tag{4.3}$$

where λ represents the wavelength of the illumination light; κ_s represents the extinction coefficient of the medium; α represents the absorption coefficient of the medium; N_0 represents the light intensity of the membrane surface; x is the distance dielectric layer through which the beam passing vertically through the beam passes. α indicates that the energy of a wavelength of light is reduced to $1/e$ of its original power over a distance of $1/\alpha$ in the medium. The reciprocal of the absorption coefficient $1/\alpha$ represents the penetration depth of the incident light in the material, which is also known as the film absorption length.

In order to obtain the total number of generated carriers, the carrier rate along the illumination direction is integrated to infinity. The carrier mobility is obtained as:

$$\mu_i = \frac{v}{|E|} = \frac{e\tau_i}{m_i} \tag{4.4}$$

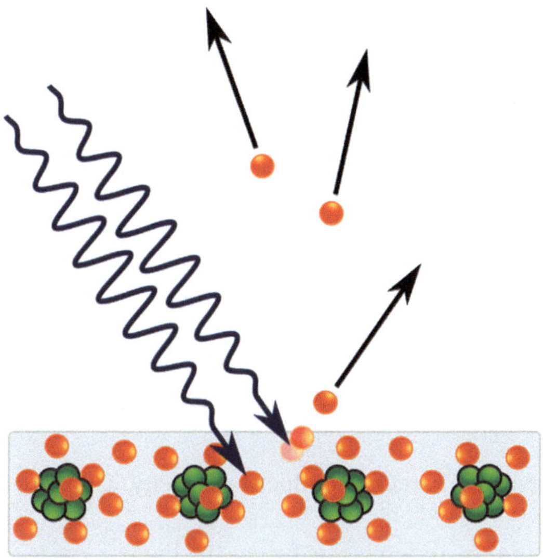

Figure 4.1 Photon absorption inducing electron emission from metal. *Source*: Ponor/ Wikimedia Commons / CC BY-SA 4.0 / https://en.wikipedia.org/wiki/Photoelectric_effect #/media/File:Photoelectric_effect_in_a_solid_-_diagram.svg / last accessed by October 03, 2025.

where v represents the carrier drift speed, e represents the electron charge, τ_i represents the carrier lifetime, and m_i represents the carrier mass.

Taking a silicon-based solar cell as an example, the silicon base is the absorbing layer. Sunlight is transmitted from the surface of the battery to the silicon base, which absorbs photons to produce electron–hole pairs and completes the separation and diffusion of charge carriers in the silicon base. In theory, the thicker the silicon base, the more photons are absorbed, i.e. the more charge carriers are produced. However, when the latter is increased to a certain thickness, the photon cannot penetrate, resulting in no carrier generation, and the carrier migration rate will be exhausted (Figure 4.1).

4.1.3 Photoelectric Conversion

The electrons in an atom have certain intervals of energy, called energy levels. Atoms in a material interact with each other to fuse energy levels into energy bands. Electron transitions between different energy bands need to overcome large energy differences. According to the tendency of the particle arrangement to obey the overall lowest energy, the electrons are thus arranged in the low-energy band. If all the energy levels are occupied by electrons, the electrons cannot move between the applied energy levels, so the full band does not conduct electricity. After all the full bands are filled, the remaining electrons gather in the highest energy band, known as the conduction band, which is the source of the material's charge carriers because of the empty energy level. All the electrons in the insulator are in the full band, and there are no electrons in the conduction band. The uppermost conduction

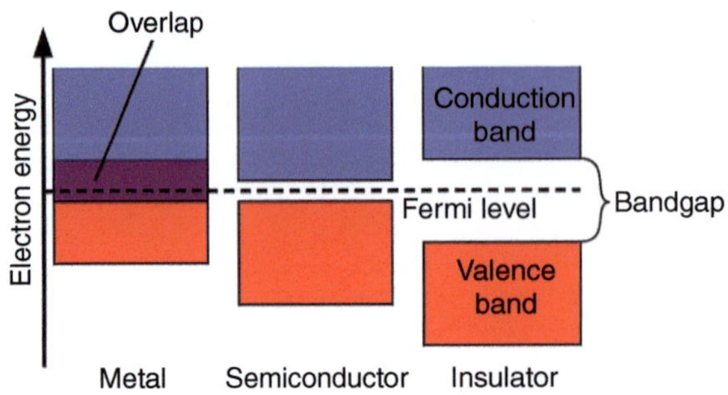

Figure 4.2 Comparison of the electronic band structures of metals, semiconductors, and insulators. *Source*: Pieter Kuiper / Wikimedia Commons / https://commons.wikimedia.org/wiki/File:Isolator-metal.svg / last accessed on October 03, 2025.

band of the metal has excess electrons and therefore conducts electricity, as shown in Figure 4.2.

The gap width E_g is used to measure the difference between the highest energy value in all energy levels in the material and the lowest energy value in the conduction band. The band gap of semiconductors is very small, ranging between 0.1 and 4 eV. The band gap of the insulator is generally higher than 4 eV. Because the bandgap of the semiconductor is relatively narrow, under some conditions, the electrons in the full band will be excited into the empty band, making it become the conduction band, leaving behind holes in the original full band. If the band gap width of the material can be excited by light energy to form an electronic transition, it is called a photoelectric material. The excitation condition is that the wavelength of the incident light cannot be greater than the wavelength limit of the material itself, i.e.

$$\lambda_0 = hc/E_g \tag{4.5}$$

where h stands for Planck's constant, and c is the speed of light in a vacuum.

Therefore, an important parameter to determine the photoconductivity effect of a semiconductor is the bandgap width, and the narrower the bandgap width, the more favorable the generation of charge carriers.

For semiconductors, according to the conductivity formula:

$$\sigma = nq\mu_n + pq\mu_p \tag{4.6}$$

where n and p represent the density of electrons and holes, respectively, μ_n and μ_p are the migration rates of electrons and holes, respectively, and q represents the quantity of electric charge.

Under the condition of illumination, when sufficient energy is provided, the photon can be absorbed by the valence band electrons in the semiconductor to produce nonequilibrium carriers, resulting in a change in the conductivity value as follows:

$$\nabla\sigma = \nabla nq\mu_n + \nabla pq\mu_p \tag{4.7}$$

$\nabla\sigma$ represents the photoconductivity.

The conductivity of the material under light conditions includes the photoconductivity and the original conductivity. The higher the conductivity under light

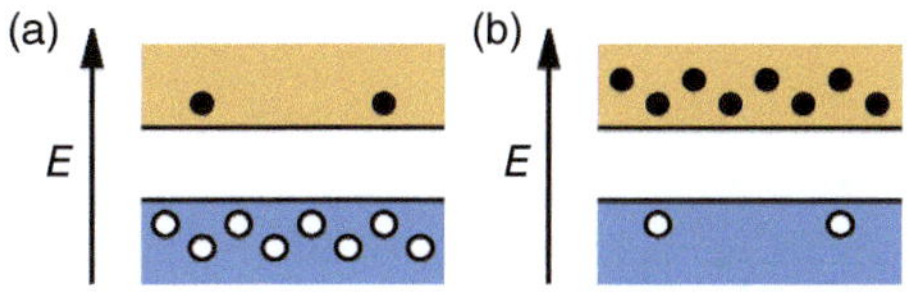

Figure 4.3 The energy band structure of (a) P-type semiconductors; (b) N-type semiconductors. Dark circles in the conduction band are electrons and light circles in the valence band are holes. *Source*: (a) Inductiveload/ Wikimedia Commons / CC BY-SA 4.0 / https://en.wikipedia.org/wiki/Extrinsic_semiconductor#/media/File:P-Type_Semiconductor_Bands.svg/ last accessed by October 03, 2025. (b) Inductiveload / Wikimedia Commons / CC BY-SA 4.0 / https://en.wikipedia.org/wiki/Extrinsic_semiconductor#/media/File:N-Type_Semiconductor_Bands.svg / last accessed by October 03, 2025.

conditions, the lower the conductivity without light, and the stronger the sensitivity of the material to light.

When a beam of light with energy equal to or greater than the semiconductor band gap is irradiated on the semiconductor photoelectric material, the electrons are excited from the valence band to the conduction band, leaving holes in the valence band, and the electrons and holes are effectively separated to realize the photoelectric conversion. When two materials with different work functions come into contact, an interaction occurs near the interface due to their different chemical potentials. Taking the photoelectric chip commonly used in OET as an example, in an N-type monocrystal silicon, since the ITO transparent conductive glass is a good conductor, electrons flow from one side of the monocrystal silicon to the ITO side until they reach an equilibrium state. At this time, the silicon is positively charged to form an electron depletion layer, and the energy band bends upward, making the silicon and ITO have the same Fermi level. The surface photovoltage is generated because the silicon surface is excited by light and generates nonequilibrium carrier, which moves in a directional way under the action of a self-built electric field, resulting in the change of surface charge. The surface barrier also changes, and the surface barrier change is the size of the surface photovoltage. If an external circuit is switched on, it forms an effective photocurrent. For P-type semiconductors, the photogenerated electrons move to the surface and the holes move to the interior. The energy band structure of P-type and N-type semiconductors is shown in Figure 4.3a and b.

Because the absorption coefficient of amorphous silicon is proportional to the light intensity and increases nonlinearly with the increase in the incident light frequency, under different spectra, amorphous silicon exhibits different responses to the electron hole pairs generated by excitation, finally leading to the change in the conductivity of amorphous silicon thin films with the incident spectra. The general dependence of the photoconductivity of amorphous silicon thin films on incident light can be described by the following relationship:

$$\Delta\sigma = \frac{e\beta\alpha\mu_n\tau_n S}{hf_l} \tag{4.8}$$

where e represents the electron charge ($e = 1.6 \times 10^{-19}$ C), α represents the absorption coefficient of amorphous silicon, β represents the number of excited electron–hole

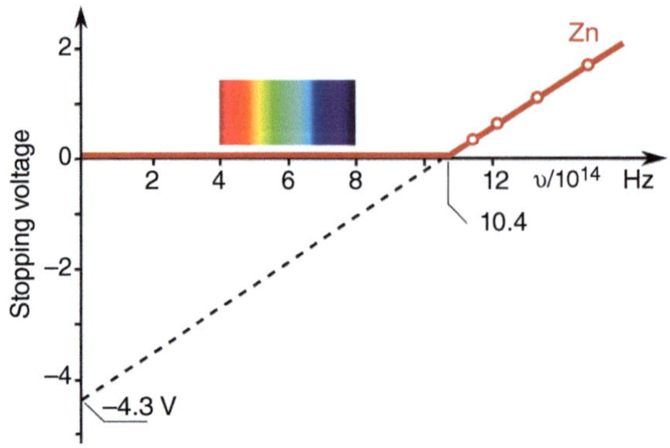

Figure 4.4 The maximum kinetic energy as a function of the frequency of light on zinc. *Source*: Ponor / Wikimedia Commons / CC BY-SA 4.0 / https://en.wikipedia.org/wiki/Photoelectric_effect#/media/File:Photoelectric_effect_-_stopping_voltage_diagram_for_zinc_-_English.svg / last accessed by October 03, 2025.

pairs per absorbed photon, μ_n represents the mobility of electrons, τ_n is the lifetime of an electron ($\mu_n\tau_n = 10^{-7}\text{cm}^2/\text{V}$), S represents the optical power, h is Planck's constant, and f_l is the optical frequency.

Therefore, assuming that all other factors in Eq. (4.8) remain constant, the change in the conductivity of the amorphous silicon film is determined. This indicates that the photoconductivity of the film will increase as the wavelength decreases, i.e. by increasing the optical frequency of the incident light on the amorphous silicon film (from red to green and then to purple), the conductivity of the film will be increased.

For the gradient electric field formed when OET is illuminated, the photoconductive material has a high impedance when there is no light, and the upper and lower electrodes form a uniform electric field. In the case of illumination, the photoconductive material excites a certain amount of charge carriers, and carrier migration leads to a decrease in conductivity, and a large amount of charge accumulates on its surface to form a nonuniform field at the spot. Therefore, in the OET system, the light intensity of the heterogeneous field is different for different band gap widths of photoelectric materials. The photoelectric excitation of materials has wavelength limitation (Figure 4.4), and the sensitivity of different photoelectric materials to the spectrum is also different.

4.2 Selection and Preparation of Photoelectric Materials

4.2.1 Selection of Photoelectric Materials in OET Chip

The vast majority of high-resistance semiconductor materials have photoconductivity effects, including common visible light region silicon (Si), hydrogenated amorphous silicon (a-Si:H), cadmium sulfide (CdS), cadmium selenide (CdSe), gallium arsenide (GaAs), indium phosphide (InP), zinc sulfide (ZnS) series in the UV region and lead sulfide (PbS), lead selenide (PbSe), indium arsenide (InAs), and indium antimonide (InSb).

From the aspects of photoconductivity, processing technology, and biocompatibility, photoinduced dielectrophoresis (DEP) chips usually require the corresponding photoconductive materials to have: (1) good character matching and thermal expansion coefficient compatibility with ITO/SnO_2/ZnO conductive glass substrates; (2) high optical absorption coefficient in the visible region; (3) ability to form a dark state conductivity matching the impedance of the water layer; (4) has a high light, dark conductance and light intensity sensitivity; (5) has the optical wavelength selectivity to meet the requirements of the device; (6) has a low transverse diffusion length; (7) doping can be well controlled to improve quantum efficiency; (8) with good process stability and low manufacturing costs; (9) insoluble in water, with stable electrochemical properties: (10) safe and nontoxic, easy to recycle.

Since visible light is often used as the lighting source in OETs, the corresponding optical absorption coefficient and light transmission distribution curve of several common photoconductive materials in the visible light band are compared. At present, the commonly used photoelectric chip materials are amorphous silicon, cadmium sulfide, and titanium oxide phthalocyanine.

4.2.2 Photoelectric Material Preparation Technology

At present, researchers mainly through physical vapor deposition, chemical vapor deposition (CVD), solution method, and magnetron sputtering prepare a variety of photoelectric thin film materials. Different preparation methods can obtain photoelectric thin film materials with different properties and structures to meet the needs of different fields, in the preparation process, to control the thickness of the film, composition and structure, and other parameters to ensure that the film material has good performance. This paper introduces the preparation method of photoelectric materials commonly used in OETs.

4.2.3 Preparation of Amorphous Silicon Materials

Amorphous silicon is a kind of highly efficient photosensitive material with a high optical absorption coefficient, and its intrinsic absorption coefficient is up to 10^6/cm. In the visible light band of 0.3–0.75 μm, its absorption coefficient is nearly an order of magnitude higher than that of monocrystalline silicon, and only about 1-μm-thick amorphous silicon film can absorb 90% of the useful visible light. Since amorphous silicon does not have the periodic atomic arrangement required by crystals, the lattice mismatch between the material and the substrate, which must be considered for crystal preparation, can be ignored. Therefore, it can be deposited on almost any substrate, including ordinary glass or silicon-based materials with transparent conductive films, and it is easy to achieve large areas. Under the condition of illumination, the electron concentration is significantly increased, and the conductivity can be increased from 10^{-10} Ω/cm to more than 10^{-5} Ω/cm, which meets the requirements for generating the DEP force: the light–dark conductivity ratio of 1000. Therefore, amorphous silicon is the photoelectric material used in most OET systems.

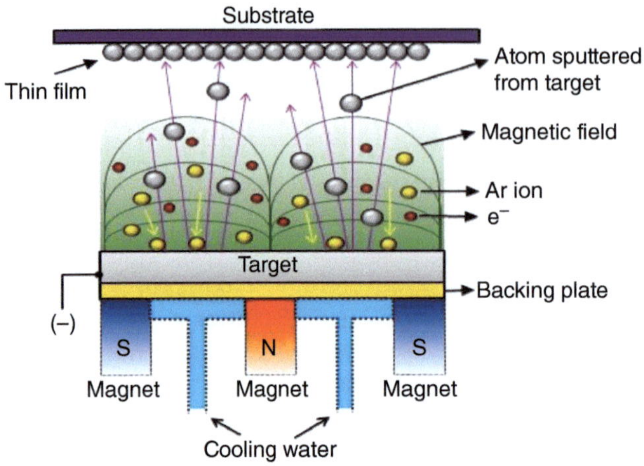

Figure 4.5 Schematic diagram of the magnetron sputtering principle.
Source: B. Janarthanan, 2021 / with permission of Elsevier.

The preparation method of amorphous silicon thin films can be divided into two categories from the preparation principle: physical method and chemical method. Common physical methods include electron beam physical vapor deposition and magnetron sputtering. Commonly used chemical methods are CVD, including plasma-enhanced chemical vapor deposition (PECVD), radiofrequency glow discharge plasma CVD, microwave plasma CVD, and so on. At the same time, there are vacuum evaporation method, DC glow discharge method, radio frequency (RF) glow discharge method, sputtering method, and hot wire method. A relatively mature process is PECVD [3]. However, the preparation condition and process cost of PECVD limit the application of the photoelectric films in OETs. In this paper, the preparation process of an amorphous silicon photoelectric film by magnetron sputtering is introduced.

The mechanism of magnetron sputtering is shown in Figure 4.5: first, Ar gas is filled into the vacuum chamber to form the required high-pressure requirements, so that Ar atoms ionize to form argon ions and high-energy electrons under high pressure. Under the double action of the magnetic field and electric field force, electrons and argon atoms will continuously collide, forming a large number of Ar ions and electrons. After one collision after another, the energy contained in the electron will be continuously reduced, and eventually it can escape the binding effect of electric and magnetic fields, and land on the anode substrate. Under the high-voltage field intensity, the cathode target absorbs a lot of energy in the impact of Ar ions on the target, which makes the target atoms break away from the surface of the target and produce new electrons. Target atoms and new electrons deposit to the anode plate to form the required conductive film under the dual action of electric field and magnetic field, and the collision probability of electrons and Ar atoms, which greatly improves the film formation rate, will be increased because the electric field and magnetic field work simultaneously and superposition.

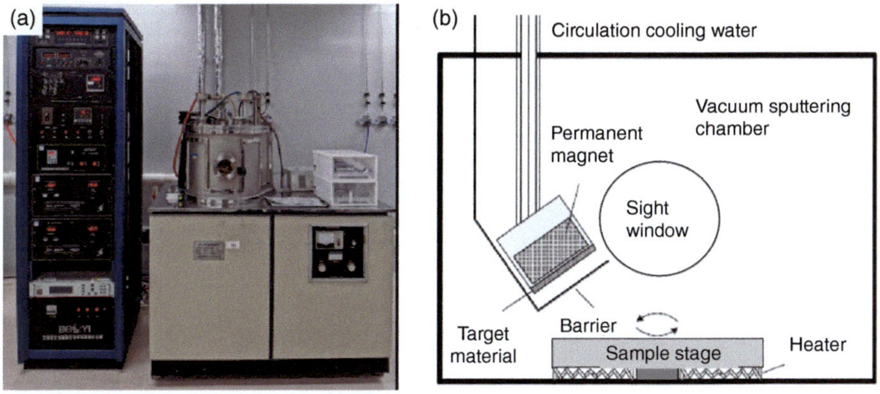

Figure 4.6 Sputtering instrument and structure. (a) Sputtering equipment system; (b) schematic diagram of the magnetron sputtering chamber.

The instrument used for magnetron sputtering for the electrooptical film plating is shown in Figure 4.6a. The equipment mainly consists of a magnetron sputtering chamber, in which there is a magnetron cathode target system, a circulating water cooling system for cooling and protecting the target and molecular pumps, a circular sample placing platform that can be rotated and equipped with a heating device at the bottom, and an observation window for observing the effect of the sputtering process, as shown in Figure 4.6b. The second is the complete vacuum system of the gas, including charging, testing, pumping three parts, used to provide sputtering auxiliary gas argon to detect the sputtering chamber pressure and achieve a vacuum environment in line with the sputtering vacuum degree; finally, the control system and power supply system include a DC power supply system for sputtering a metal film and a radio frequency power supply device that can be coated with a nonmetallic film on the substrate or a metal film on it, and all motion control is a pneumatic valve composed of a mechanical mechanism composed of ordinary gas in the cylinder to inflate and pump to drive the piston movement, so as to realize the switch control of the valve.

The specific process of the photoelectric film preparation process by a sputtering equipment is as follows. The process of the microfluidic chip prepared by magnetron sputtering is shown in Figure 4.7:

Amorphous silicon films with different thicknesses prepared by magnetron sputtering are shown in Figure 4.8, with thicknesses of 500 nm, 1000 nm, and 1500 nm, respectively. The greater the thickness, the lower the transparency of the film, and the darker the color.

4.2.4 Preparation of a Cadmium Sulfide Photoelectric Film

The main performance parameter of the photoelectric material used in OET is the light–dark conductivity ratio, and some materials can be selected by referring to the materials used in the photoresistor. The dark resistance of the photoresistor is generally above 1.5 MΩ, the carrier rising high light resistance can reach 1 kΩ

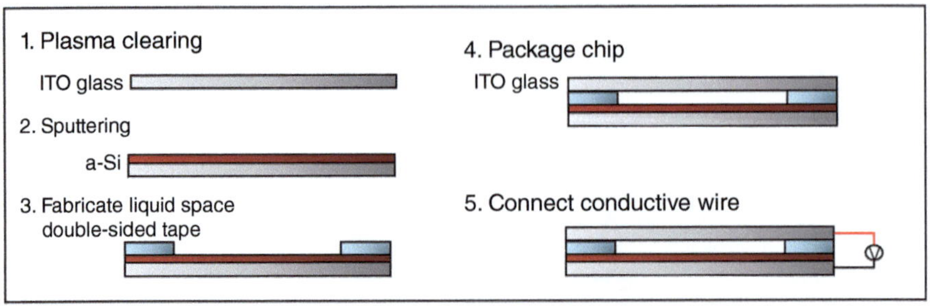

Figure 4.7 Preparation process of a microfluidic chip combined with magnetron sputtering.

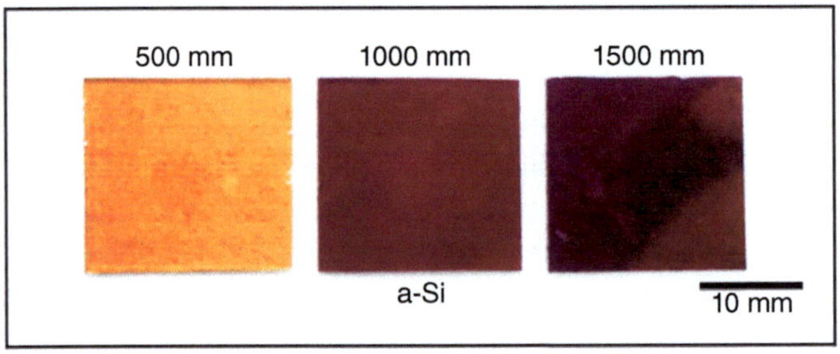

Figure 4.8 Preparation of amorphous silicon photoelectric films with different film thicknesses by magnetron sputtering.

under light, and the light–dark conductivity ratio can reach 1500. The photoresistor material is commonly CdS. CdS, as an N-type semiconductor material, has a band-gap width of 2.45 eV, which has a good absorption effect in the visible near-UV region, and serves as a buffer layer in solar cells, which is a key layer between copper indium gallium selenide (CIGS) and ZnO. CdS can be used as an N-type semiconductor material, and CIGS and ZnO band gap gradients can also be adjusted. In order to prepare transparent CdS films with high resistivity, the common methods include CVD, electroplating, RF magnetron sputtering, and chemical water bath. This paper introduces the principle of electroplating and electroless plating for the preparation of CdS photoelectric films.

4.2.4.1 Electrochemical Plating

The plating process of a CdS film is as follows:

$$S_2O_3^{2-} + 6H^+ + 4e^- = 2S + H_2O \tag{4.9}$$

$$S + H^+ + 2e^- = HS^- \tag{4.10}$$

$$Cd^{2+} + HS^- = CdS + H^+ \tag{4.11}$$

The process steps are as follows: prepare the drugs needed for the experiment; calculate the mass or volume required for each drug; build the electroplating platform;

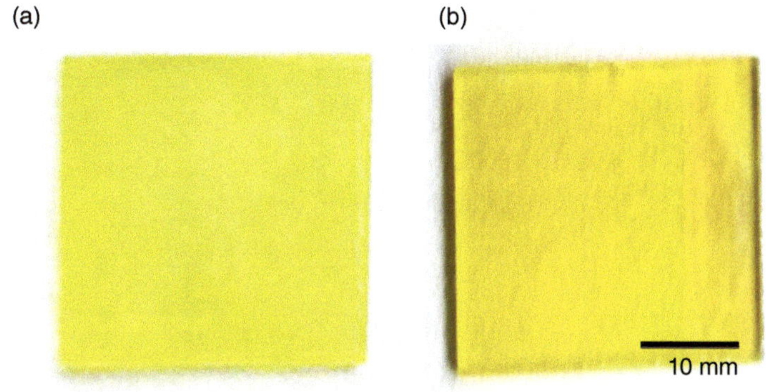

Figure 4.9 Cadmium sulfide film electroless plating. (a) A single coating; (b) secondary coating [4].

heat the water bath to the desired temperature (80° C); add the configured plating solution to the plating bath; control the electroplating voltage and current, and finally, the film is annealed.

4.2.4.2 Chemical Plating

Electroless plating is based on the principle of redox reaction on the surface of the activated matrix coated with a layer of metal coating with excellent mechanical and conductive properties (Figure 4.9a and b). Compared with electroplating, chemical and physical vapor deposition methods, electroless plating has the advantages of low-cost, uniform coating, and can be plated on the surface of metals, semiconductors, and nonmetals. Its biggest feature is that it can obtain uniform and dense coating on the surface of a complex matrix such as deep holes and profiles.

The reaction equation of electroless plating is as follows:

$$NH_4^+ + OH^- = NH_3 + H_2O \tag{4.12}$$

$$Cd^{2+} + 4NH_3 = Cd(NH_3)_4^{2+} \tag{4.13}$$

$$Cd^{2+} + 2OH^- = Cd(OH)_2 \tag{4.14}$$

$$(NH_2)_2CS + OH^- = SH^- + H_2O + +H_2CN_2 \tag{4.15}$$

$$SH^- + OH^- = H_2O + S^{2-} \tag{4.16}$$

$$Cd^{2+} + S^{2-} = CdS \tag{4.17}$$

4.2.5 Preparation of TiOPc

Titano-phthalocyanine (TiOPc) is a kind of organic small molecule semiconductor material in metal phthalocyanine derivatives, which is easy to use as a photoelectric

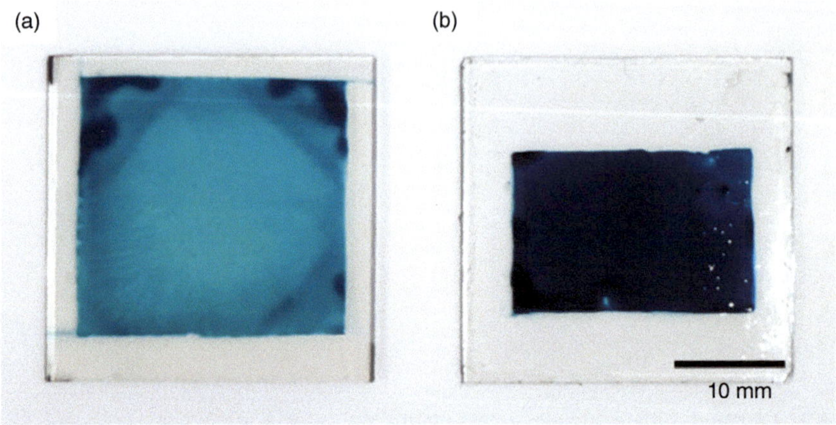

Figure 4.10 TiOPc spin-coated film. (a) Spinning coating applied on ITO; (b) final product.

material for flexible thin films [5]. The material has high photosensitivity in the visible and near infrared regions, and it has good photoelectric properties, with an electrical conductivity of 10^{-12} to 10^{-8} S/cm. It has been widely used in photodetectors, photocells, organic light-emitting diodes (OLEDs), organic transistors and other fields. TiOPc films can be prepared by the spin-coating method.

The process of preparing a TiOPc thin film by spinning coating is as follows: prepare the reagent and proportionally allocate suspension; ultrasonic dispersion of the suspension; the obtained spin-coating liquid is filtered and coated on ITO (Figure 4.10a), and the rotation speed and time are controlled. And place on the hot drying plate to dry, and finally spin-coating the protective layer (Figure 4.10b).

4.2.6 Micro-well Array Photolithography

In the process of fabricating micro-well array OET chips, the preparation of micro-well arrays requires the photolithography technology. Photolithography is a core process in the semiconductor manufacturing field, specifically referring to the process where photosensitive resist (photoresist) undergoes photodecomposition or photopolymerization reactions under exposure to high-intensity light sources, transferring complex patterns from a mask onto semiconductor materials such as silicon substrates (Figure 4.11).

As shown in Figure 4.12, the photolithography operation on the substrate of an amorphous silicon-ITO-glass mainly consists of several steps, including the substrate surface treatment, photoresist spin-coating, pre-baking, alignment and exposure, post-baking, developing, and observation under a microscope [6].

The substrate surface treatment mainly consists of three parts: ultrasonic surface cleaning, high-temperature baking treatment, and plasma activation treatment.

Figure 4.11 The spin coater equipment used in the experiment (left) and the URE 2000S photolithography machine (right).

In the ultrasonic cleaner, the substrate is cleaned with ethanol and water sequentially, and the surface liquid is dried with a nitrogen gas gun. Subsequently, the substrate is subjected to a high-temperature treatment at 120° C in an oven for 120 minutes. After removal, the substrate is allowed to cool to room temperature and then placed in a plasma machine for the activation treatment. The above substrate surface treatment alters the chemical properties of the amorphous silicon surface, increases surface roughness, and enhances the adhesion of the photoresist on the amorphous silicon substrate.

Next, the photoresist is uniformly spin-coated on the centrifugal spin coater. After spin-coating the photoresist, the substrate is pre-baked on a constant-temperature heating plate for 30 minutes to evaporate most of the solvents in the photoresist. After cooling to room temperature, the substrate is aligned and exposed for 18 seconds in the middle position using a URE 2000S contact photolithography machine with a micro-well array mask. The exposed substrate is then post-baked for 5 minutes. After cooling to room temperature, the substrate is developed with a propylene glycol methyl ether acetate (PGMEA) developer for 2 minutes, followed by rinsing with deionized water to remove the developer, drying the surface moisture, and observing the lithography effect under a microscope.

After the fabrication of the micro-well array, the bonding of the polydimethylsiloxane (PDMS)-glass and microchannel design and cutting are required (Figure 4.13), followed by chip encapsulation to complete the process.

Finally, the previously fabricated micro-well array silicon wafer, PDMS-glass-ITO substrate, and double-sided adhesive microchannels (Figure 4.14) are aligned based on the positions of the microchannel inlet and outlet holes. Then, conductive wires are attached to the two ITO surfaces, and the chip is fixed onto a carrier glass slide, completing the encapsulation and fabrication of the OET micro-well array microfluidic chip (Figure 4.15).

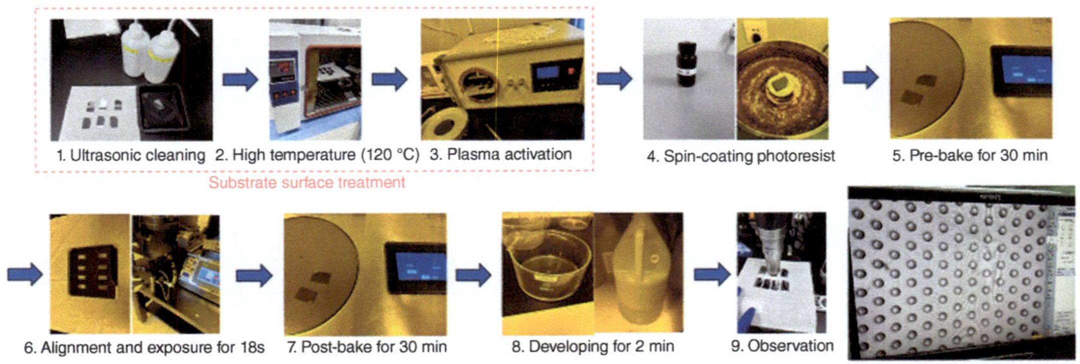

Figure 4.12 Photolithography operation steps.

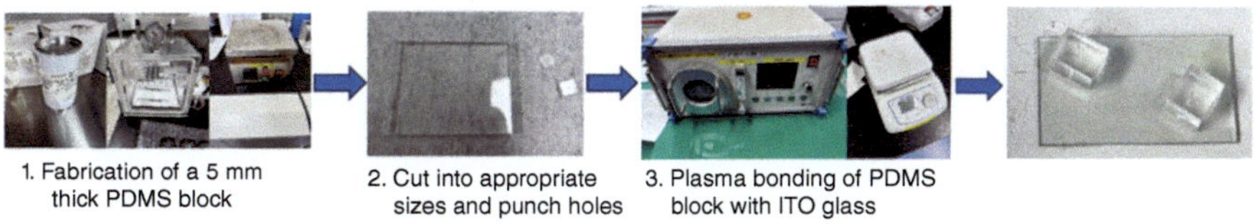

Figure 4.13 OET microfluidic chip bonding steps.

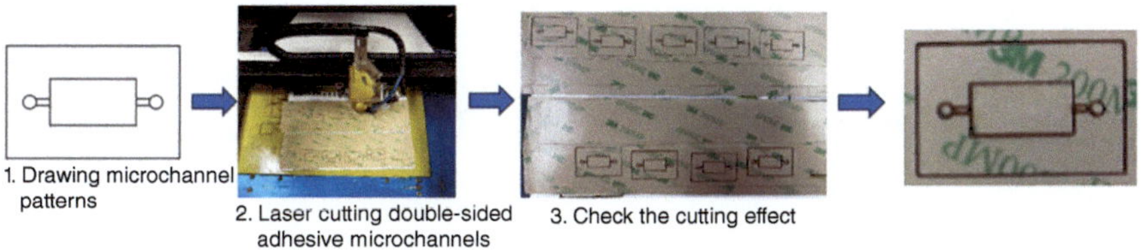

Figure 4.14 Microchannel fabrication steps.

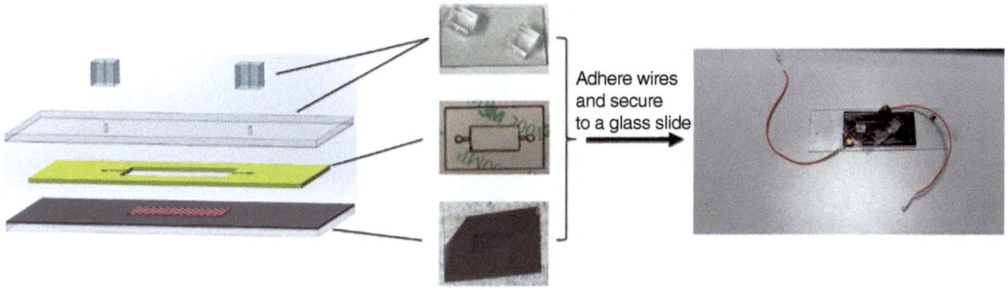

Figure 4.15 Chip encapsulation.

4.2.7 Characterization of Photoelectric Film Properties

After the preparation of the photoelectric film, it is necessary to test the performance of the photoelectric film to evaluate the quality of the film and select the photoelectric film with high stability, good uniformity, high carrier mobility, and good film forming quality to improve the performance of the microfluidic chip.

4.2.7.1 Structure and Morphology Characterization

During the characterization of the surface morphology of the sample, the surface feature image was collected by a scanning electron microscope and the thickness of the film was measured by a step meter (Figure 4.16a and b).

Scanning electron microscopy (SEM) is one of the main methods to test the surface morphology of materials. SEM bombards the surface of the sample with an extremely narrow electron beam and scans it point by point. The physical signals generated by the interaction between the electron beam and the sample, such as secondary electrons, backscattered electrons, auger electrons, and transmitted electrons, are received by the detector, amplified and converted into modulated signals, and finally the images with the characteristics of the sample surface are displayed on the display.

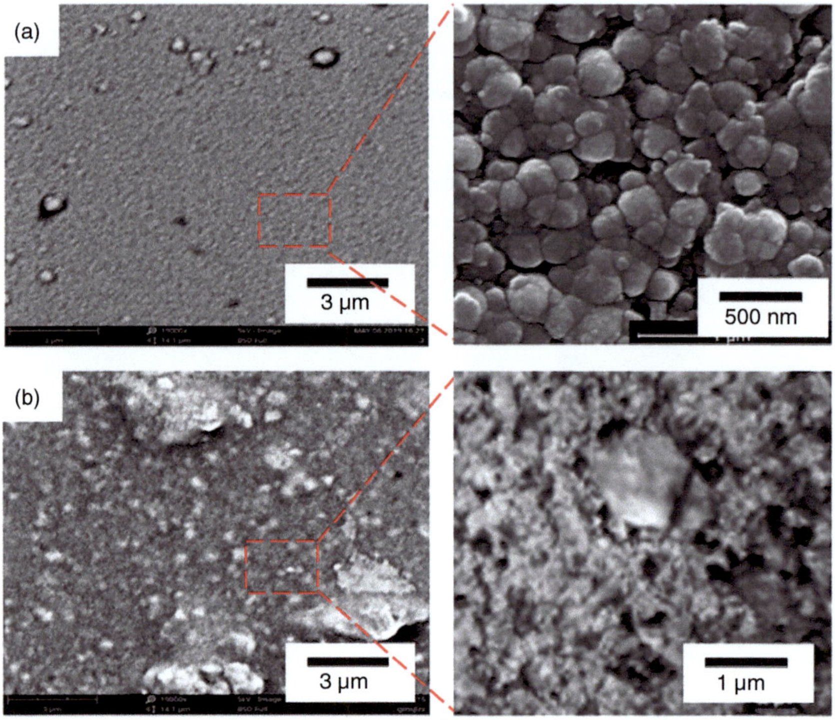

Figure 4.16 Scanning electron microscopy of the photoelectronic film. (a) Cadmium sulfide film; (b) titanium phthalocyanine oxide film.

A step meter is a probe contact measuring instrument used to characterize the surface topography of materials. The measurement principle is that when the probe slides from the surface of the measured film, it moves up and down along the tiny peaks and valleys on the surface. The sensor output electrical signal after the measurement bridge output amplitude modulation signal, this signal is proportional to the displacement of the probe. The amplitude–modulation signal is amplified and processed by phase-sensitive rectification, and the amplified slow-changing signal is obtained. Therefore, the surface profile of the measured film can be reflected by the motion path of the probe, but the step meter can only reflect the tiny peak and valley profile of the linear distance of the film surface.

4.2.7.2 Optical Characterization

In order to characterize the optical properties of the film, the crystal structure of the film was analyzed by X-ray diffraction (XRD) (Figure 4.17a–d), and the atomic composition and component content of the film were obtained by X-ray photoelectron spectroscopy (XPS) (Figure 4.18a and b). The absorptivity and light transmittance of the film were measured by a UV–visible spectrophotometer.

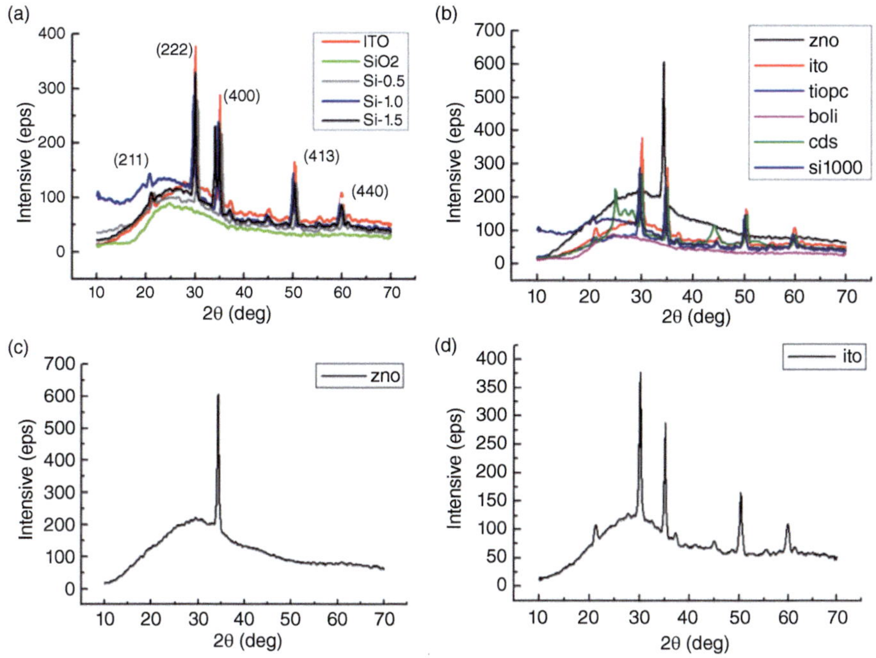

Figure 4.17 (a) XRD of amorphous silicon films of different thicknesses; (b) XRD of cadmium sulfide and titanium phthalocyanine oxide films; (c) standard reference cards for zinc oxide films; (d) standard reference card for conductive glass ITO film.

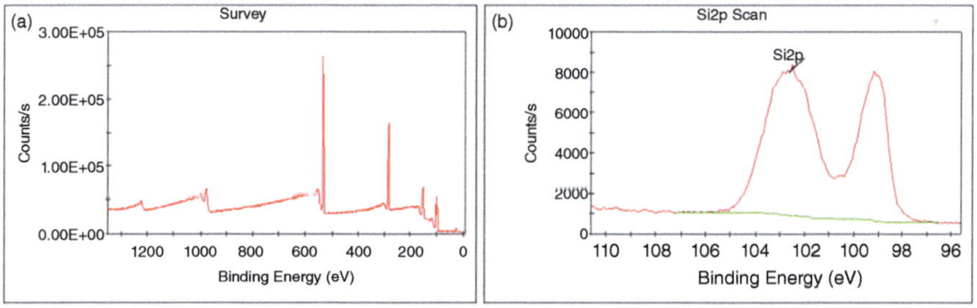

Figure 4.18 XPS electron spectra of amorphous silicon films. (a) Energy spectrum of amorphous silicon films; (b) energy spectrum of silicon in thin films.

An X-ray diffractometer is a kind of instrument used for accurate phase analysis, and the qualitative analysis and quantitative analysis of measured substances by the principle of diffraction [7]. In 1912, the German physicist Max von Laue proposed and experimentally confirmed that when X-rays are irradiated to the crystal, they are scattered by atoms on all sides of the crystal, and the atoms produce their own scattered waves and interfere with each other, and eventually diffraction occurs. The superposition of diffraction waves makes the intensity of the rays stronger in

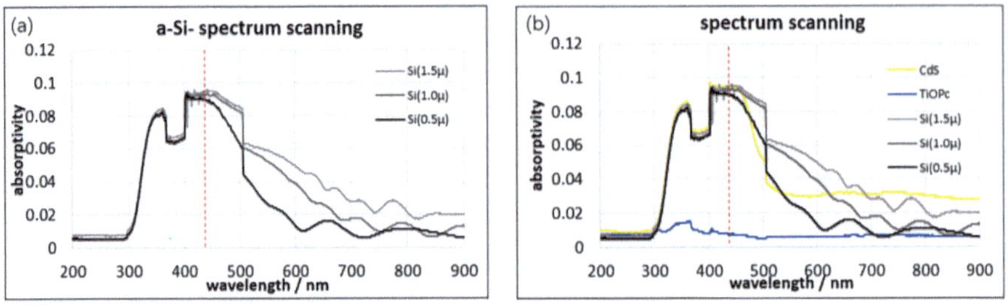

Figure 4.19 The absorption rate of the film was measured by a UV-visible spectrophotometer. (a) Absorption rate of amorphous silicon films of different thicknesses; (b) comparison of visible light absorption rate of three kinds of photoelectric films.

some directions and weaker in some directions. The crystal structure can be obtained by analyzing the diffraction results.

An X-ray photoelectron spectrometer is an advanced instrument used for surface analysis mainly to characterize the surface elements of materials and their chemical states [8]. It can provide information on the composition and content of elements, molecular structure, valence states, chemical bonds, etc. based on the overall, surface, micro-region, and depth distribution of materials. On the other hand, XPS incident with a photon beam effectively reduces the damage to the sample, which is very conducive to the characterization and analysis of polymer materials and organic materials.

The spectral absorption of amorphous silicon was analyzed using an UV spectrometer, which showed that a-Si films had excellent absorption in the wavelength range of 300–550 nm, as shown in Figure 4.19a and b. It indicates that the light intensity gradient should be increased to produce a stronger electric field [4].

4.2.7.3 Electrical Properties

In the characterization of the electrical properties of the prepared photoelectric thin films, the resistance, voltage and current of the materials are often measured by the coplanar method (Figure 4.20). Under different conditions of light and no light, the light to dark ratio of the film can be calculated by measuring the resistance of the film respectively (Figure 4.21a and b).

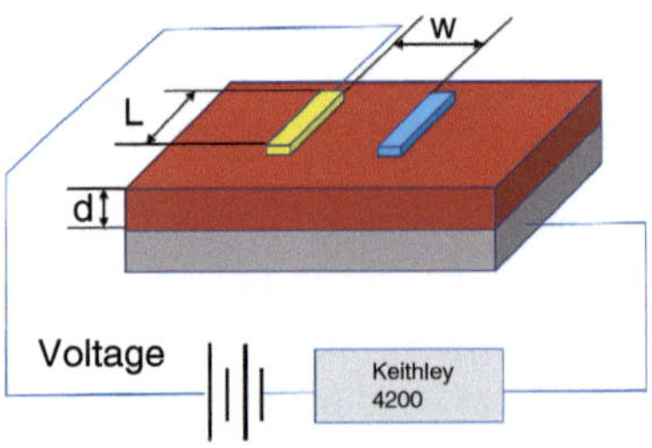

Figure 4.20 The resistivity of the photoelectric thin film was measured by the coplanar method.

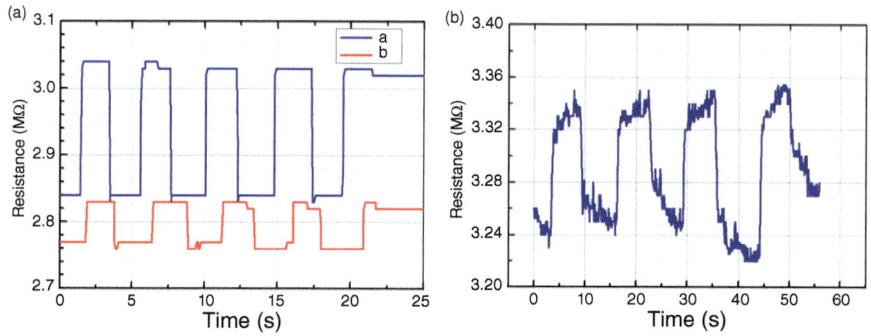

Figure 4.21 Resistance test of the photoelectric film. (a) Testing of amorphous silicon films; (b) cadmium sulfide film test.

The measured resistance R of the film can be calculated by the formula:

$$R = \frac{\rho W}{S} = \frac{U}{I} \tag{4.18}$$

where ρ represents the resistivity of the material, W represents the conductor length of the resistance, S represents the cross-sectional area of the resistance wire ($S = d \times L$), U represents the voltage at both ends of the resistor, and I represents the current passing through the resistance. The conductivity is the reciprocal of the resistance:

$$\sigma = \frac{1}{\rho} \tag{4.19}$$

The relation between the conductance G (in Siemens, S) and the resistance R (in ohms) is $G = 1/R$.

References

1 Cheng, T.-P. (2013). *Einstein's Physics: Atoms, Quanta, and Relativity-Derived, Explained, and Appraised*. Oxford UP.

2 Smith, W. (1873). Effect of light on selenium during the passage of an electric current. *Nature* 7 (173): 303–303.

3 Stalin, B., Ravichandran, M., Jasper, S., and Kannan, C.R. (2020). Synthesis and characterization of brass–AlN composites synthesized by ball milling. *Materials Today Proceedings* 22: 2573–2581.

4 Liang, S.Z. (2020). *Based study on Microorganism Robot Control via Multifunctional Photoelectronic Tweezers System*. Beijing, China: Beihang University.

5 Chen, W., Huang, L., Qiao, X. et al. (2012). Efficient planar organic solar cells with the high near-infrared response. *Organic Electronics* 13 (6): 1086–1091.

6 Cheng, X.T. (2024). *Research on Fluorescence Microsphere Detection Technology Based on Microwell Array Optoelectronic Tweezers Chip*. Beijing, China: Beihang University.

7 Matsunami, N., Kakiuchida, H., Sataka, M., and Okayasu, S. (2013). *XRD Characterization of AlN Thin Films Prepared by Reactive RF-sputter Deposition.* Scientific Research Publishing.

8 Sharma, N., Ilango, S., Dash, S., and Tyagi, A. (2017). X-ray photoelectron spectroscopy studies on AlN thin films grown by ion beam sputtering in reactive assistance of N+/N2+ ions: substrate temperature induced compositional variations. *Thin Solid Films* 636: 626–633.

5

Experimental Techniques and Setups

Understanding the intricacies of the OET setup and operation is crucial for maximizing their potential in various applications. This chapter briefly introduces the basic structure of three types of OET devices and provides a detailed description of the hardware and software components of a classical system. It then discusses the overall construction of the OET system and presents an advanced product, highlighting its functionalities and features. Finally, the chapter summarizes common challenges in using OET devices and suggests potential solutions.

5.1 Overview of Experimental Setup

OET systems consist of multiple key components. A typical OET setup includes a signal generator, a digital micromirror device (DMD) projector, an upright microscope, a computer, and an OET chip. The DMD projector, controlled by a computer interface, generates light patterns that are then projected through the microscope's objective onto the OET device's photoconductive layer. The observation light path combines the light pattern and particle cells, which are captured by a charge-coupled device (CCD) camera connected to the microscope for imaging. The function generator supplies an alternating current (AC) signal to operate the OET device. The OET chip serves as the system's control core, responsible for positioning particles and cells and responding to light pattern changes to create a non-uniform electric field for particle manipulation.

There are several configurations for light projection in OET systems [1]. One model involves directing light from above the device, sharing the objective lens for both illuminating the sample and observing it, as depicted in Figure 5.1a [2]. Another model projects light from below, opposite the observational path, allowing for different analytical perspectives (Figure 5.1b) [3]. These systems are adaptable to various photoconductive materials. For example, devices utilizing hydrogenated amorphous silicon (a-Si:H), titanyl phthalocyanine (TiOPc), or bulk heterojunction (BHJ) polymers on indium tin oxide (ITO)-coated glass substrates, all of which are semi-transparent to visible light, can operate with either projection setup. However, phototransistor-based OET devices require the top-down projection model due to their opaque silicon substrate, which inhibits light passage from

Optoelectronic Tweezers for Manipulation at the Micro and Nano Scale, First Edition.
Lin Feng, Menglu Tan, Ao Wang, and Jiaying Zhang.
© 2026 WILEY-VCH GmbH. All rights reserved, including rights for text and data mining and training of artificial intelligence technologies or similar technologies. Published 2026 by WILEY-VCH GmbH.

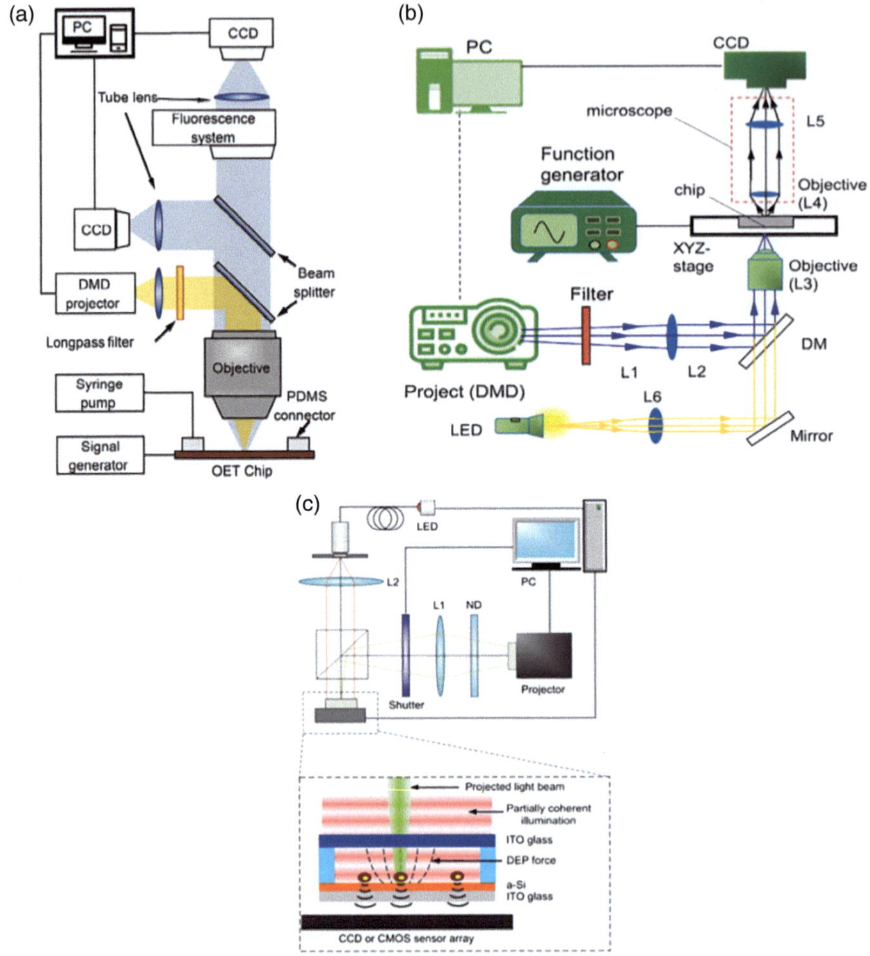

Figure 5.1 (a) Diagram of an OET system where light patterns are projected from above the device. (b) Diagram of an OET system with light patterns projected from beneath the device. (c) Diagram of an OET system utilizing a lens-free holographic microscope for setup. *Source*: (a) Gan, et al., [2] / John Wiley & Sons. (b) Liang, et al., [3] / Frontiers Media S.A. / CC BY 4.0. (c) Kuo-Wei, et al., 2013 / Royal Society of Chemistry.

below. This adaptability and versatility in projection methods highlight the flexibility of OET systems in different experimental and research contexts.

A crucial component of the OET setup is the light source. OET operation does not necessitate coherent light sources and can function effectively with a low optical density, leading to the utilization of various light sources in the OET research. Commonly employed light sources for OET platforms include DMD projectors and liquid-crystal display (LCD) projectors. Additionally, self-assembled optical systems comprising a DMD chip and light sources such as light-emitting diodes (LEDs), lasers, and Hg lamps have been utilized in OET studies [4]. These light sources play a crucial role by acting as optical displays that project dynamic light patterns onto the surface of the OET device. This projection forms "light-activated virtual electrodes," which are integral to the system's programmable manipulation capabilities. By varying the light patterns, users can actively control and modify the electric field distribution on the device's surface, enabling the precise and customizable manipulation of particles.

Figure 5.1c showcases a streamlined version of the OET system that utilizes a lens-free holographic microscope [5]. This innovative setup allows for the effective manipulation of numerous individual cells and microparticles over a substantial field of view, extending up to 240 mm^2. The broad reach of this system makes it particularly adept for handling large numbers of particles simultaneously, improving efficiency in various experimental contexts.

In addition to its primary function of micromanipulation, the OET experimental framework is versatile enough to integrate additional functionalities like detection and fabrication capabilities. For instance, systems employing laser-based Raman spectroscopy [6] can be incorporated to provide molecular-level insights during manipulation processes. Similarly, UV patterning systems [7] can be paired with OET devices for fabricating intricate patterns at microscales. Moreover, integrating microfluidic systems [8] with OET platforms can enhance applications related to fluid dynamics and improve precision in controlling biological assays or chemical reactions. These extensions not only expand the utility of OET systems across various scientific investigations but also demonstrate their adaptability and potential for fostering advancements in fields such as biotechnology, materials science, and analytical chemistry.

5.2 Detailed Description of the Optoelectronic Tweezer Device

A common OET device is described here. The OET system consists of two parts: hardware and software. The hardware part is designed based on the principle of OET, including optical path analysis, simulation verification, and equipment construction; the software part achieves high efficiency and accuracy of operation through a remote-control platform and an automated control program. This section focuses on the design and construction process of the OET control system, which uses optoelectronic microfluidic chips to generate dielectrophoretic force to control particles and combines with a cloud platform to achieve the remote control. Through this system, experimenters can perform the real-time control and image feedback on mobile terminals, breaking the limitations of experimental space and environment, and providing higher flexibility and convenience for various micro-manipulation experiments.

5.2.1 Optical Path Design

In terms of hardware implementation, OET basically requires a projection pattern generation optical path, an illumination optical path and an experimental observation optical path, which are usually built using a microscope, a projector, and an optical condensation system. Combined with the existing commercial microscope system, the illumination optical path and the experimental imaging observation optical path can be simplified, mainly to achieve the combination of the condensed beam imaging of the projection optical path.

1) *Projection optical path*: It is composed of a projector based on a DMD chip and a condensation lens system, which images the reconfigurable light pattern projected by the projector onto the OET chip.
2) *Illumination optical path*: It is composed of a light source and an objective lens, which is mainly used to facilitate the observation and identification of target particles on the chip.
3) *Observation optical path*: It is composed of a CCD camera and an objective lens, which images the object on the OET onto the CCD and forms a real-time photo of the particles and projection pattern on the chip that can be observed on the computer. The illumination optical path and the observation optical path can actually be realized with the help of a standard microscope, so the main design link of this system lies in the design of the projection optical path.

Because the core component of the entire OET system imaging is the DMD on the projector, the imaging path parameters of the entire projection optical path should be designed according to the size characteristics of the DMD. Among them, DMD is an array composed of multiple high-speed digital light reflection switches, which is composed of many small aluminum reflective mirrors. The number of lenses is determined by the display resolution, and a small lens corresponds to one pixel. The projector images the object on the DMD device. Through the pixel-level controllable characteristics of the DMD device and its high-speed flip frequency, each image point is scanned onto the detector in turn, realizing the high-speed passive point scanning imaging of the object under visible light conditions. Active scanning imaging can also be achieved by adding appropriate light sources. The projector and DMD form the front part of the optical path of the projection pattern, as shown in Figure 5.2.

After controlling the particle size comparison, a DMD chip with an imaging pixel of 1080*960 pixels is adopted. The equivalent object image size in the optical path is an image with a screen ratio of 4:3 and a diagonal length of 0.55 in. The standard size of the projector imaging used is 78 in., the actual focal length is between 22.04 and 24.14 mm, and the projection ratio is 1.96–2.15, which changes with the focal length. According to the lens imaging principle (as shown in Figure 5.3):

$$\beta = \frac{y'}{y} = -\frac{f}{x} = -\frac{x'}{f'} \tag{5.1}$$

where β is the magnification. Then, at the minimum focal length of the projector:

$$\beta = -\frac{78}{0.55} = -141.82 \tag{5.2}$$

$$\text{ratio} = 1.96,\ f' = 22.04\ \text{mm},\ x' = 3125.67\ \text{mm} \tag{5.3}$$

$$l' = x' + f' = 3127.71\ \text{mm} \tag{5.4}$$

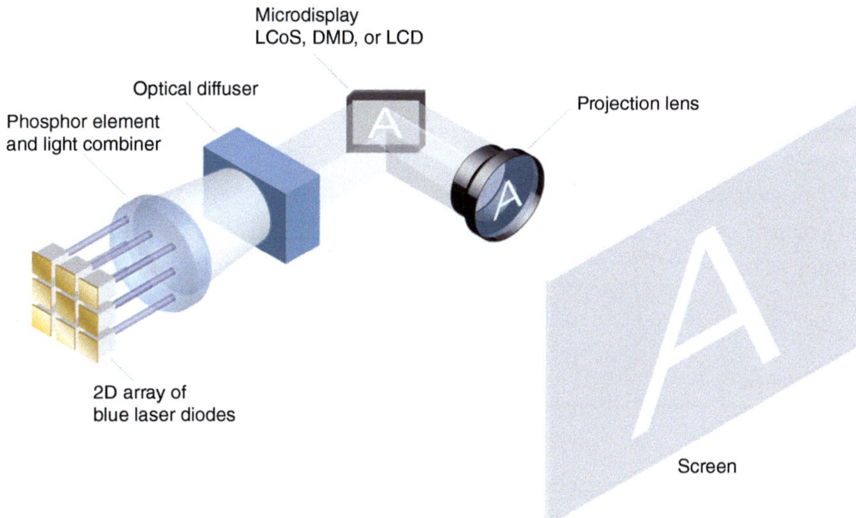

Figure 5.2 Projector imaging optical model. *Source*: Laser Illuminated Projector Association / Wikimedia Common / CC BY-SA 4.0 / https://commons.wikimedia.org/wiki/File:Laser-blue-animation.gif / last accessed by October 03, 2025.

Figure 5.3 Lens imaging principle.

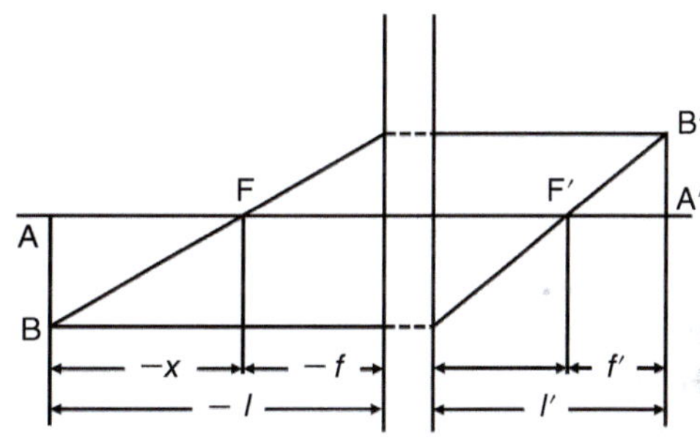

At the projector's maximum focal length:

$$\text{ratio} = 2.15,\ f' = 24.14\ \text{mm},\ x' = 3423.49\ \text{mm} \tag{5.5}$$

$$l' = x' + f' = 3447.63\ \text{mm} \tag{5.6}$$

Therefore, when the projection screen is 78 in., the imaging distance of the projector is adjusted between 3127.71 and 3447.63 mm by fine-tuning the focal length f' between 22.04 and 24.14 mm. The change in the focal length causes a slight change in the distance between the equivalent lens and the object. To facilitate the subsequent adjustment of the imaging position and fine-tuning of the instrument, the average projection focal length is taken as the reference focal length of the optical path. The parameters of the projector at this time are shown in Table 5.1.

And the angle of the pattern from the projector lens is:

$$\theta = 2\arctan\left(\frac{y'}{l'}\right) = 33.56^\circ \tag{5.7}$$

Table 5.1 Projector equivalent optical model data.

	Minimum focal length	Intermediate focal length	Maximum focal length
Screen size		78 in. = 1.9812 m	
Vertical magnification β		−141.818	
Focal length f/f'	22.04	23	24.14
Projection distance l'	3127.71	3284.81	3447.63
x'	3125.67	3261.81	3423.49
x	0.1554	0.1622	0.1702
Object distance	22.20	23.16	23.31

The front part of the projection optical path is composed of a projector containing a DMD, and the rear part of the optical path needs to scale the optical pattern projected by the projector, and finally focus it to a size that matches the microfluidic chip through an objective lens. First, a lens is needed to scale the image projected by the projector, and then a lens is used to convert the optical path into a parallel optical path, and then a group of lenses is used to shrink the optical path to a size consistent with the optical aperture of the projection objective.

In addition, for the illumination optical path, the light generated by the light source can be focused into parallel light through the fourth convex lens, and then matched to a size consistent with the optical aperture of the projection objective through a lens group, and finally projected onto the optoelectronic microfluidic chip through the objective lens to achieve the illumination requirements of the object. For the real-time display observation optical path, the image of the optoelectronic microfluidic chip needs to pass through the objective lens to become a parallel optical path, and then a group of lenses is used to scale the optical path size to a size consistent with the optical sensing effective area of the image sensor CCD. The final designed OET optical system is shown in Figure 5.4.

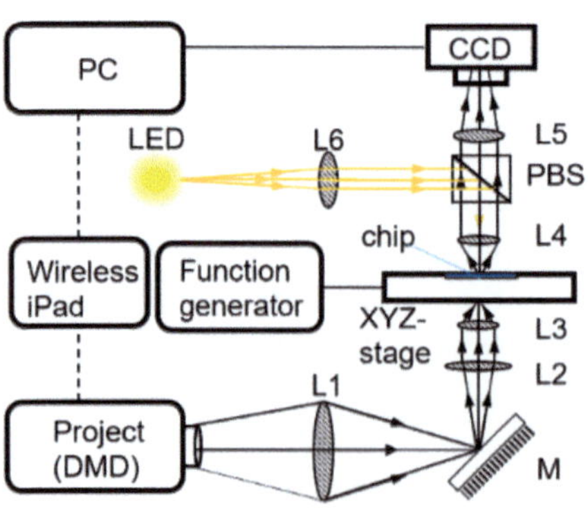

Figure 5.4 Design of the optical system of OET. *Source*: Shuzhang, et al., 2021 / MDPI / CC BY 4.0.

5.2.2 Optical Path Simulation

In order to analyze the parameters of the functional lenses such as image zooming and imaging of the optical path of the OET system, the designed optical path is analyzed equivalently with the help of an optical simulation software.

For the rear part of the projection optical path, the first lens selection needs to adjust the equivalent lens of the projector to a larger and farther image, and to a smaller and indented image. When the angle of the projector's emitted light is 33.56° and the distance between the lens and the projector lens is 15–30 mm, the diameter of the light landing on the lens is 23–32 mm. For the maximum distance of 30 mm, the reduced imaging is 58 mm. Therefore, lens1 with a diameter $D = 40$ mm and a focal length $f1 = 100$ mm is used for preliminary analysis. The distance to the projector is 23 mm. Since all the emitted light rays of the edge phase point are in the direction away from the X-axis, the phase point close to the edge will not be captured by the next lens. When the focal length is reduced, the size of the lens will be limited, and the reduction in size will cause more serious image loss. Finally, a lens with a diameter of 30 mm and a focal length of 60 mm is used.

In order to further scale the image optical path, the second lens is placed at the imaging point to further converge the light. Therefore, the parameters of the second lens are diameter $D = 40$ mm and focal length $f2 = 100$ mm.

After the action of the first two lenses, the light has been controllably converged to a position close to the optical axis. In order to scale the image optical path to a size that matches the aperture of the objective lens, a lens group is used to scale the light beam in parallel. Therefore, the parameters of the third and fourth lenses are diameter $D = 20$ mm, and focal lengths $f3 = 50$ mm and $f4 = 30$ mm, respectively.

The design data is imported into the ZEMAX optical design software for optical simulation analysis, as shown in Figure 5.5.

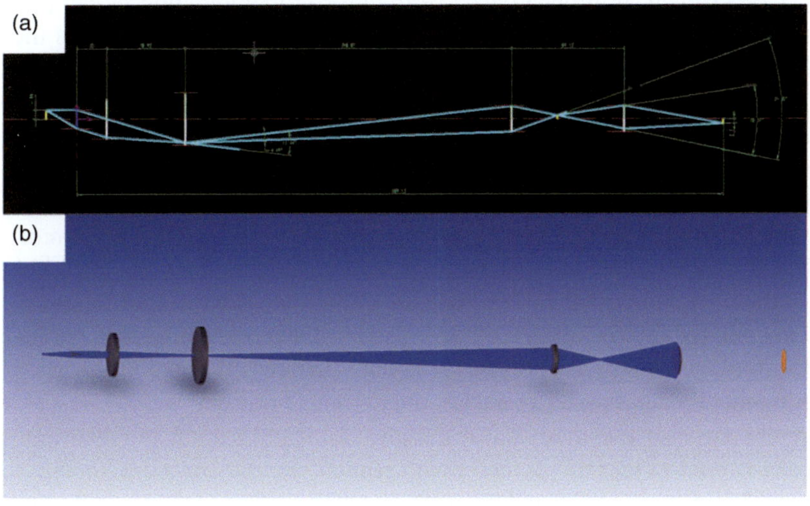

Figure 5.5 Equivalent simulation of the projection optical path. (a) Equivalent optical path; (b) three-dimensional effect of equivalent optical path.

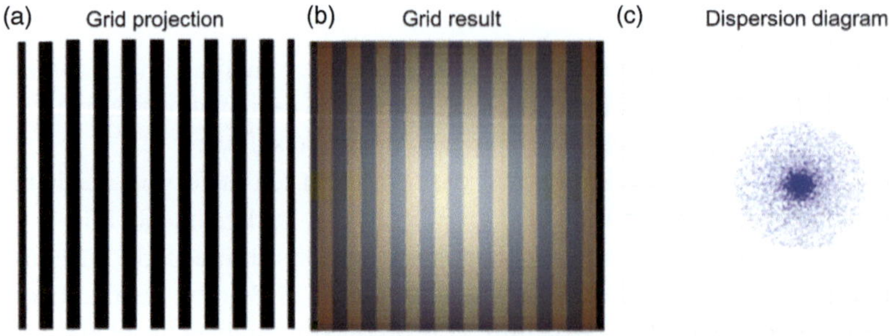

Figure 5.6 Simulation of the optical path imaging effect. (a) Grid test image; (b) grid test result; (c) optical path dispersion distribution.

The simulation shows that under the wavelength of visible light, the imaging point accurately falls on the designed focal plane position, which is consistent with the design result. Image simulation is used to analyze the imaging effect of the OET projection optical path. By setting a black and white grid image (Figure 5.6a), the contrast and distortion of the actual optical imaging are tested. It is easy to observe when the grid imaging effect is biased. The simulation imaging analysis results are shown in Figure 5.6b. Due to the inevitable spherical aberration and coma in the preparation of the lens, the refraction effect produced when different wavelengths of light are irradiated is different, causing the dispersion of the system, as shown in Figure 5.6c. However, its light and dark contrast has met the basic requirements for the operation of OET.

5.2.3 Experimental Control Strategy Design

In the OET experiment, in order to achieve the precise manipulation of cell particles, a feedback-based control strategy was adopted. This strategy combines the image processing technology with the proportional-integral-derivative (PID) control algorithm to ensure the high-precision movement of cell particles in complex environments.

Before the experiment begins, the light pattern needs to be corrected first. Since the image captured by the CCD camera will have barrel distortion at the corners, we use a projector to project a chessboard-shaped optical pattern at the four corners of the CCD camera field of view. Through the built-in functions in Opens, the position of the projection point is detected and associated with the coordinates in the projector, and the interpolation method is used to obtain the complete mapping coordinates from the camera coordinates to the DMD, thereby eliminating distortion and ensuring that the light pattern matches the CCD field of view.

For different particles, two methods are designed for identification and detection. The first one is used to identify particles with regular circular contours, in which key contour features are extracted through a preprocessing process including graying, blurring, and binarization, and the coordinates are determined using the Hough circle transform. In addition, the gray pattern matching method is

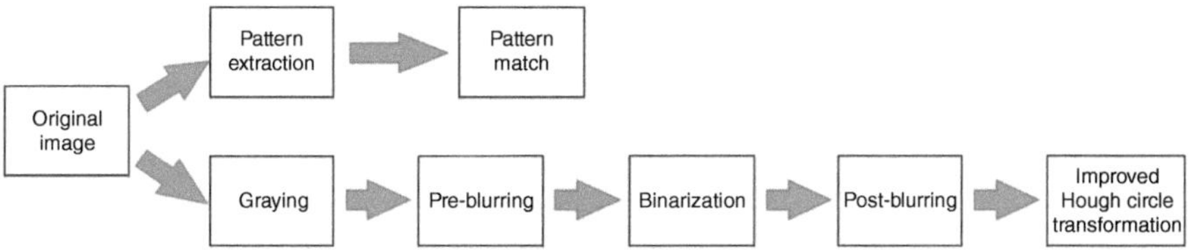

Figure 5.7 Preprocessing flow of video image acquisition.

used to process irregular particles. The specific image recognition process is as follows:

1) Video Frame Acquisition
Through the mature and convenient video acquisition method of OpenCV, the frames of the OET system control video files and the real-time camera images are acquired.
2) Video Frame Preprocessing
Before identifying the obtained cells, the image is preprocessed such as noise reduction to match the corresponding cell recognition function (Figure 5.7). The final recognition effect depends largely on the quality of the preprocessing, and different preprocessing methods are used for different application scenarios. Therefore, under the same processing function name, the function parameters are fine-tuned to adapt to the actual processing situation.
Among them, the most critical and effective processing method is image binarization, i.e. the acquired image is divided into foreground and background. Binarization uses a certain grayscale value as the segmentation standard, which is called the threshold. Currently, the maximum interclass variance method (Otsu's algorithm) is usually used to calculate a reasonable segmentation threshold.
In the Otsu's algorithm, the threshold is set to t, w_0 is the foreground area ratio, u_0 is the average grayscale, w_1 is the background area ratio, and u_1 is the average grayscale. The average grayscale value of the image is:

$$u = w_0 u_0 + w_1 u_1 \tag{5.8}$$

$$g = w_0 (u_0 - u)^2 + w_1 (u_1 - u)^2 \tag{5.9}$$

Finally, the maximum value T corresponding to t is obtained by calculating the interclass variance g, and the segmentation threshold of the collected image is obtained.
3) Image Cell Particle Recognition
After completing the image preprocessing, the operation object needs to be effectively identified to realize the information extraction and analysis of the particles, and finally complete the feedback of the system and automatic micro-operation. For circular particles, the Hough circle transform method is used to identify circular particles. Hough circle transform: Generate a cumulative coordinate plane based on a potential point on a circle that may correspond to a nonzero pixel

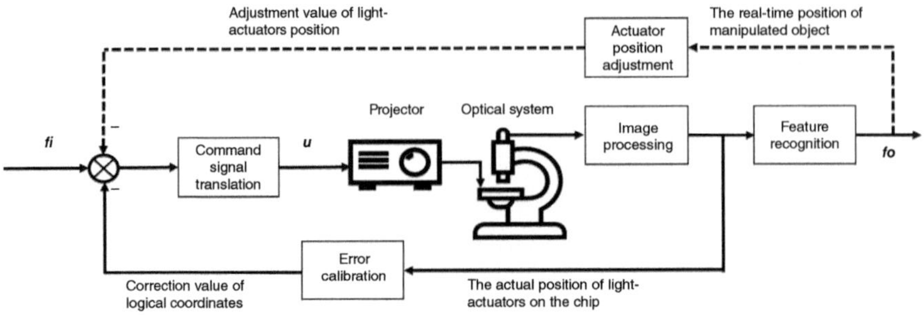

Figure 5.8 Negative feedback automatic control system model. *Source*: Shuzhang, et al., 2021 / MDPI / CC BY 4.0.

point in the image, and locate the circle by judging the cumulative weight by the maximum value.

For noncircular particles, the template matching method is used for identification. Template matching: By traversing all pixels of the image, calculating the similarity between the image block and the defined template, and locating similar positions using the similarity judgment standard, the non-circular target can be identified.

An image-based closed-loop control system is used to improve the operation accuracy and realize partial automatic manipulation. As shown in Figure 5.8, the structure of the closed-loop control system is shown, where the input signal fi represents the logical coordinates of the optical actuator used to control the target object, and the output signal fo provides the actual coordinates of the target object on the chip. The control system applies two main negative feedbacks. The first is the error calibration feedback, which aims to eliminate the difference between the coordinates of the object in the logic and chip system, and the real-time error is monitored by the CCD camera on the microscope. On the other hand, the error introduced by the hardware setting is systematic and can be compensated by negative feedback calibration. The dotted line in the figure indicates that the feedback is only used in the automatic operation mode, and the spot moves the target particle along a specific trajectory and adjusts its position in real-time to match the moving target position [9].

5.2.4 System Construction

In order to test the optoelectronic film in the OET system and conduct particle manipulation experiments, the prepared optoelectronic film is used to construct an optoelectronic microfluidic chip [9], as shown in Figure 5.9. The chip consists of a bottom electrode of the optoelectronic film, an intermediate dielectric solution layer, and a top ITO electrode. The bottom photosensitive electrode is a 1 μm sputtered amorphous silicon film on a 200-nm-thick ITO film, and then a 150-μm-thick particle manipulation cavity is built on the photosensitive electrode using a double-sided tape. Finally, an ITO conductive glass (a resistance of 10 Ω, a size of 25 mm × 25 mm) is used as the top electrode. The upper and lower electrodes

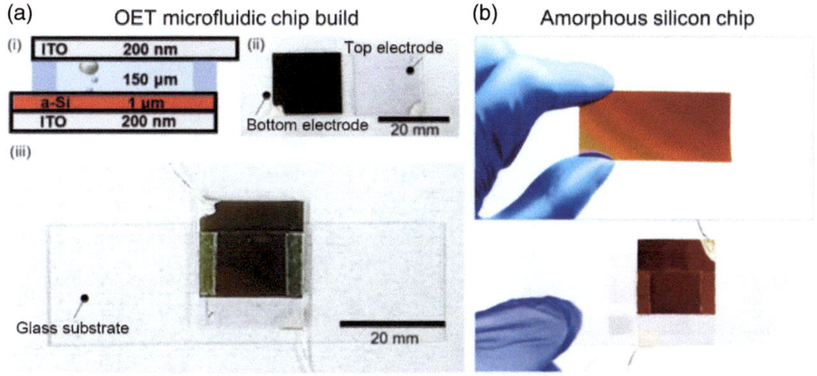

Figure 5.9 Construction of the OET microfluidic chip. (a) Schematic and composition of the microfluidic chip structure; (b) amorphous silicon chip.

are connected by conductive silver paste for the control of external power signals. In order to prevent the optoelectronic film from being oxidized and cracked, the conductive silver paste is cured in a vacuum drying oven at 120° C. (The experimental details can be found in [10–14].)

Based on the above simulation analysis results and the calculation of optical path propagation and imaging effects, corresponding optical lenses and other optical components were purchased. The actual system construction is based on four modules: illumination optical path, projection control optical path, observation imaging optical path, and input signal source. The projection control optical path is built based on the DMD projector. The projection pattern is reduced to match the control scale of the optoelectronic microfluidic chip through lenses (L1 and L2), focusing lenses (L3) and reflectors, and the projected image of the projector is controlled by computer programming. The particle manipulation imaging is observed in real-time with the help of a 20*x* objective lens (L4), lens (L5), and a CCD camera. An LED light is used as a straight beam to focus the illumination through a focusing lens (L6). The function generator outputs an AC voltage and applies it to the top ITO electrode and the bottom photoconductive layer ITO electrode of the optoelectronic microfluidic chip of the OET. The final construction of the OET platform is shown in Figure 5.10.

Based on the hardware and software architecture design of the OET system, the module composition of the system, the collaboration mode of the hardware, the functional distribution of the software, and the development mode were designed and constructed. In order to remotely control the OET system, a remote control platform based on a cloud database was established. The remote control platform includes a cloud server, a multi-platform interactive terminal, and an execution terminal containing the OET device. The cloud server is a commercial server that can handle a large amount of time-consuming computing load, including image processing, projection graphics generation, and data format conversion. The interactive terminal includes a cross-platform graphical user interface (GUI) implemented on iPad and computer PC, which supports effective interaction between the operator and the OET system. The execution terminal is a computer

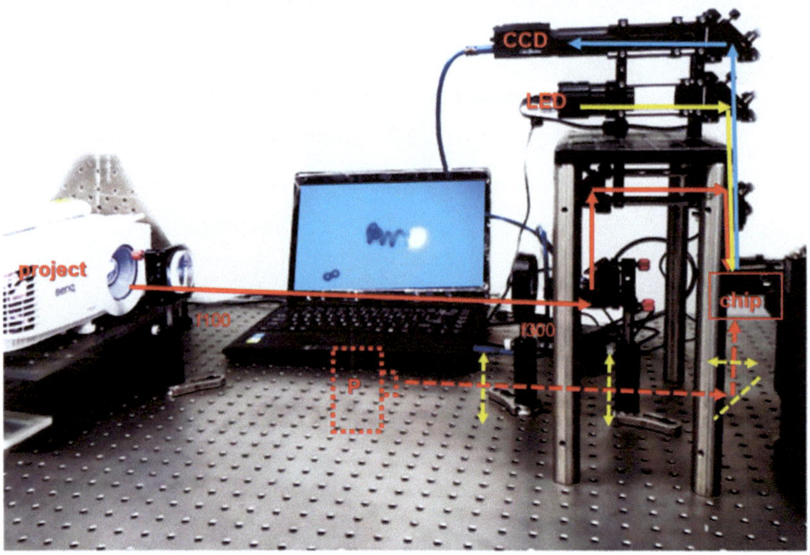

Figure 5.10 Construction of the optical system of OET.

connected to the microscope and projector, responsible for sending control signals to the experimental equipment and receiving result feedback. These three modules are deployed under the wireless LAN and use the WebSocket protocol for secure cloud communication.

The cloud computing-based platform can be divided into the following structures: display module, middle layer module, and bottom device module. The display layer is the program interaction interface, which realizes the human-computer interaction and directly affects the overall quality of the cloud platform. The middleware layer is the background service program in the traditional sense, which plays a role of inheriting the past and ushering in the future, providing image recognition services and resource storage services, etc. The infrastructure layer is the hardware equipment in the traditional sense, which provides the corresponding computer hardware service resources for the middleware layer. The more common methods are public cloud and private cloud.

Based on the optical path system of OET, according to project requirements and laboratory conditions, the cloud server is selected as Mac Pro, which also serves as iOS development equipment, the interactive device is selected as iPad and ordinary PC, and the execution terminal is selected as the PC attached to the instrument purchased by the laboratory. All devices are deployed in the local area network so that they can transmit data by wire or wireless through the transmission control protocol/internet protocol (TCP/IP) protocol. The PC equipped with the test equipment is used as the execution terminal to connect with the experimental equipment to ensure compatibility with the test equipment. Placing all devices in a local area network can bring the best security. It is difficult for external connections to directly access any internal device.

The execution module directly connects the OET device and the computer to obtain the real-time image of the FlyCapture camera and draw the OET operation graphics. By putting the entire thread of acquiring the image into a new thread, the efficiency of acquiring the image is improved, and the program is prevented from blocking the GUI thread during the acquisition, processing, and sending of the image.

In order to make the image and other data transmission rate of the negative feedback system faster and reduce the response time. By using WebSocket and Json protocols, full two-way channel communication is realized under a single TCP connection, avoiding the use of HTTP protocol. By using the multi-threading technology, the functions of saving and sending images are placed in non-main threads for execution, reducing the impact of long-term command execution. In the data transmission process, the acquired image data needs to be converted into the Json format for fast network transmission. The received image data is restored to the original QImage format so that other functions and programs can complete the call and further processing of the image data. First, the user sends a link to the server through the handshake request of the HTTP protocol. After the handshake is successful, a two-way data transmission connection is established. Based on this link, real-time two-way data transmission is realized. Finally, the whole transmission process is ended through a close request initiated by negotiation between the two parties.

For the user execution interaction module, a standard GUI program and interface are designed for convenient control of the OET system, as shown in Figure 5.11. It integrates the real-time CCD camera monitoring, design, and control of projection graphics, post-processing of collected microscope images, and display of

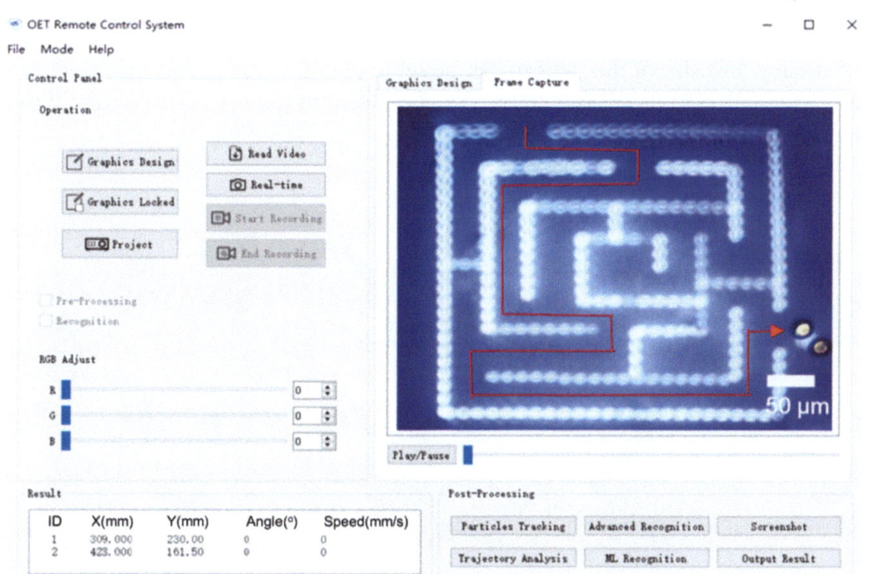

Figure 5.11 Human-computer interaction interface for remote control platform operation.

recognition results. The projected graphics are designed using iPad GUI and sent to the execution terminal computer through a remote server. In the OET control GUI, the mode menu is designed as manual operation (autonomous drawing to operate the OET), semi-automatic operation (controlling existing OET graphics to complete specific tasks), and fully automatic operation (the system autonomously identifies particles and automatically generates pre-programmed OET graphics and completes specific operation tasks) to meet the control requirements of different control targets. Drawing control buttons and video control buttons are set in the operation selection area to realize real-time video processing, such as image preprocessing, image recognition, and image RGB adjustment.

In addition, the QPaint module and QGraphicsView module in the Qt software library are used to provide a drawing system for the control pattern. At the same time, with the help of the mouse event control, the mouse start and end positions are recorded to complete the drawing of the graphics. The graphics drawn by QPaint are transferred to the QGraphicsView framework in the form of Pixmap, combining the two functions so that users can freely draw the required experimental manipulation patterns.

By performing complex centralized data processing in high-performance servers, the efficiency and professionalism of the system can be improved, and remote cross-platform operation can be achieved. Two modes of the remote wireless control system operation via iPad are shown in Figure 5.12a and b. Based on the fixed function mode of OET, there are two main components in the interface, which are used for the grabbing and handling of particles such as cells and collection chambers. Among them, the opening and closing of the ring is achieved by setting the "O" and "C" buttons, and the enrichment of cell particles in the collection chamber is achieved by "M" and "C." For the free control module, the corresponding particle manipulation image is provided by freely drawing rectangular lines, and the color, transparency, and size of the pattern can be set to obtain the required manipulation light wavelength and light intensity to avoid damage to the biological tissue of the manipulated cells.

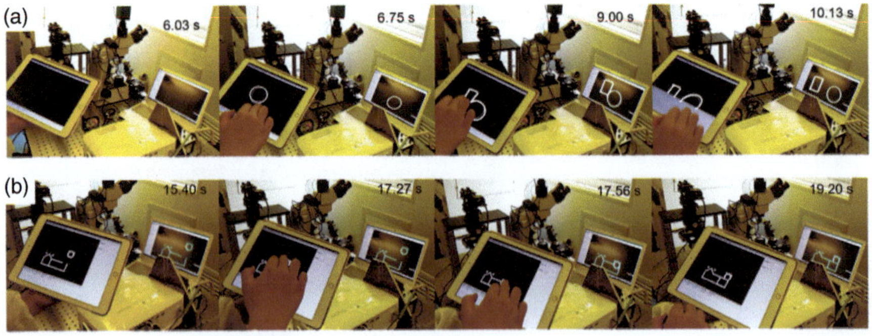

Figure 5.12 Demonstration of the remote wireless control of the OET platform. (a) Free drawing control mode; (b) Fixed control mode.

5.3 Advanced OET Product Introduction

Light Operator S1 is a self-developed innovative OET micro/nano manipulation platform by a Micro Nano Robot [15], primarily achieving the following functions:

1) Microspheres (Figure 5.13)
   - Micro-patterning and micro/nano manipulation
   - Based on the size difference separation, separate individual microspheres from a single microsphere suspension, or sort and recover specific types of microspheres.
2) Cells (Figure 5.14–5.16)
   - Single-cell manipulation and cell clustering manipulation
   - Interactions between individual cells and single microspheres, as well as multiple target objects
   - Cell movement path planning and parallel cell manipulation
   - Organoid manipulation and sorting

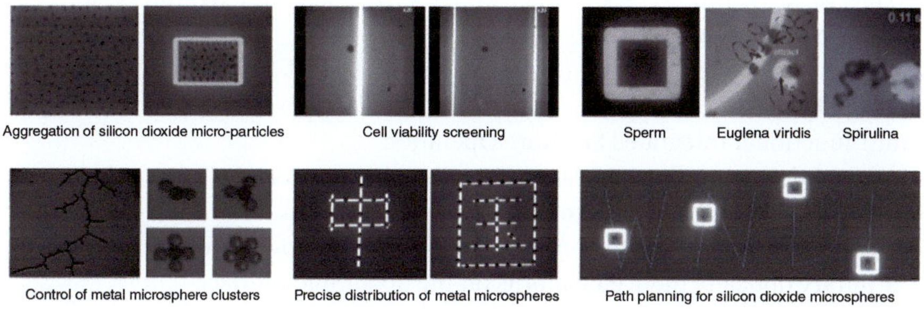

Figure 5.13 Control of different microspheres.

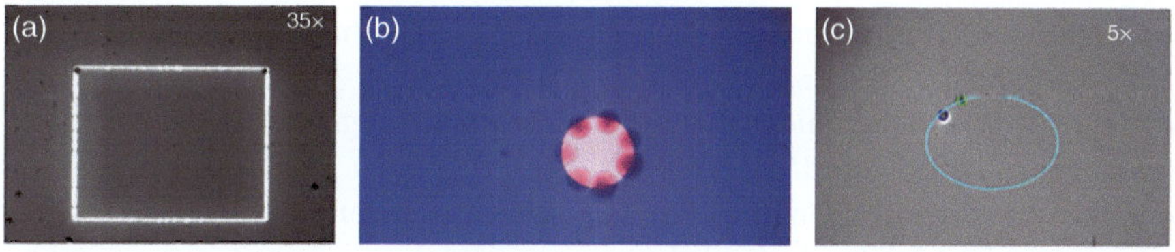

Figure 5.14 (a) Cell viability screening (b) Hydrogel microsphere-loaded stem cell growth observation experiment (c) Cell path planning.

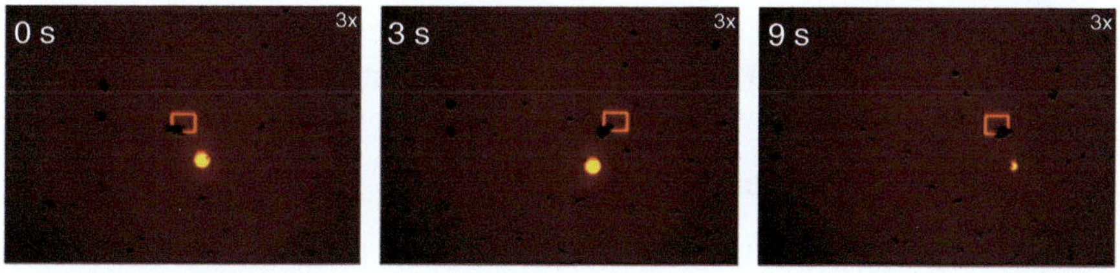

Figure 5.15 Multidimensional assembly of cell microparticles in a hydrogel viscous liquid.

Figure 5.16 Organoid manipulation and sorting.

- Single-cell screening and cell sorting, separating individual cells from a single-cell suspension, or sorting specific target cell subpopulations, and achieving live cell recovery

The innovative features of this product include:

1) High Operational Throughput
   - The number of parallel operating objects can exceed 500
   - The fastest operating speed of microspheres can reach over 1000 μm/s, and cell speed can reach over 50 μm/s
   - Can use the dielectric properties, size, and biological activity of the manipulated objects for screening
2) Multi-functional Integrated Software Operation
   - Automatic recognition of target objects based on models
   - Multi-field image stitching and shadow compensation
   - User-defined aperture element control algorithm, allowing for the creation of arbitrary optically induced patterns on the operating interface
   - Supports animated image import, element combination, and secondary development functions
3) Optional multi-module functionality embedding
   - Temperature control module for the precise temperature control of the chip from 4 to 50 °C
   - Magnetic control module platform for adding opto-magnetic coupling fields to magnetically controlled target objects
   - Fluorescence control module for bright-field observation and three types of fluorescence imaging bright-field observation
4) Strong functional expansibility
   - Supports chip design for different channels and different experimental targets (Figure 5.17a-c)
   - Can be integrated with a microscope system
   - Can be integrated with Raman spectroscopy systems and mass spectrometry detection systems, etc.

Introduction to the operating software

The Light Operator S1 operating software is designed to easily complete instrument operation settings, run monitoring, target manipulation, and image and data export. This software is responsible for controlling instrument operations, from sample

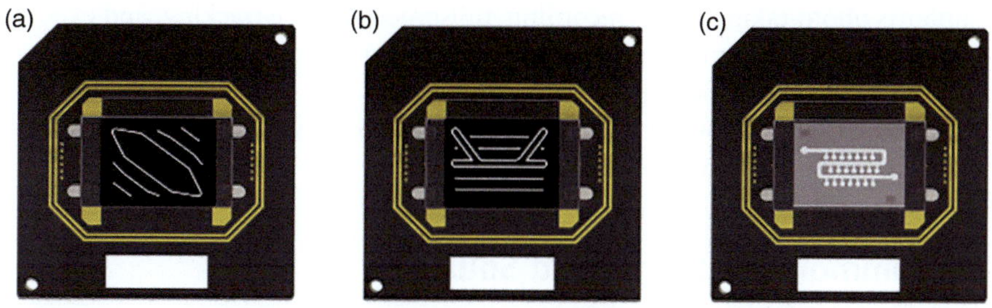

Figure 5.17 (a) A simple-to-operate chip prepared with amorphous silicon as the photosensitive layer, convenient for scientific research experiments. Easy to disassemble and clean, ready to use as soon as it is disassembled. (b) The chip includes microchannels, and solutions are pumped into it for use. The channels can be used for path planning, and the chip can recover experimental solution samples. (c) The chip supports custom channels, making it convenient for complex and diverse application scenarios.

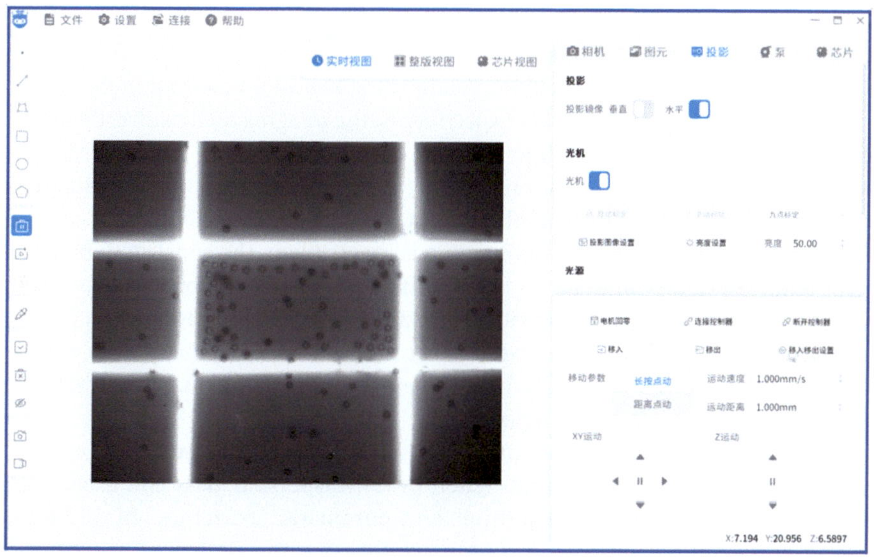

Figure 5.18 Software operation interface.

import, manipulation, sorting to exporting target cells. The software operation interface is shown in Figure 5.18. It offers the following functions:

1) Direct control of the stage, light source, automatic pump, and other hardware
2) Built-in image AI algorithm for automatic target object recognition
3) Supports custom aperture, multi-field image stitching, and shadow compensation functions for clear images
4) Supports multiple graphic drawing, element combination, path planning, and animated image import
5) Standard four-channel triple fluorescence, customizable fluorescence channels, fluorescence, and bright-field switching

6) Supports photo-taking, video recording, full-process image tracking, and export
7) Simple and convenient operation interface, allowing users to complete experiments with simple clicks
8) Supports setting user permissions to ensure data security
9) Supports secondary development

5.4 Common Challenges and Solutions

5.4.1 Hardware System

In optical experiments and applications, the stability of the hardware system directly affects the quality of the experimental results. In particular, the performance of the optical imaging system determines the clarity, resolution, and accuracy of the image. In this process, the stability of the optical path system, the distortion control of optical imaging, the stability of the pattern (such as no flicker), and the guarantee of resolution are the core challenges. In addition, the consistency of the chip is also a key factor in ensuring the performance of the optical system.

The stability of the optical imaging system is the basis for ensuring high-quality images. Especially in high-precision experiments, the imaging quality directly determines the reliability of the experimental data. During the imaging process, the path of the light beam, the focus position, and the adjustment accuracy of the imaging instrument may lead to distortion, blur or instability of the image.

The distortion problem of optical imaging is inevitable in the construction of the optical path, especially in systems using complex optical components. Common types of distortion include geometric distortion, chromatic aberration, lens distortion, etc. These distortions can cause unclear images and even affect the accuracy of experimental results. Especially in high-resolution imaging, distortion is often more obvious and has a greater impact.

In order to ensure that the imaging system is distortion-free, researchers usually use high-quality optical components (such as low-dispersion glass and aspherical lenses), which can minimize distortion and chromatic aberration. In addition, numerical calculation methods are used to optimize the design of the optical system, and the high-order optical compensation technology is used to correct the distortion problem in imaging to ensure that each optical element in the optical path can achieve the best imaging effect. By introducing the distortion correction algorithms (such as the image reconstruction technology and optical compensation algorithms), the distortion in the imaging system can be further reduced and the image quality can be improved.

In dynamic optical imaging and image capture, the stability and consistency of the light pattern are problems that must be solved during the experiment, especially in high-speed imaging and image capture systems. The flicker of the light pattern is often caused by factors such as the instability of the light source, the mismatch of the sampling frequency of the image sensor, or the vibration of the optical components. This light pattern flickering phenomenon will interfere with the imaging effect and even affect the accuracy of data acquisition.

The key to ensuring the stability of the light pattern is to control the stability of the light source and the frequency synchronization of the system. The use of a high-stability light source (such as a laser diode or a stable LED light source) can avoid the flicker problem caused by light source fluctuations. In addition, the image acquisition system should have a high-frequency sampling capability to ensure that the sampling of the imaging sensor is synchronized with the light source frequency to avoid the occurrence of flicker. In system design, reducing external interference through vibration isolation technology and ensuring the stability of optical components and sensors is also an effective way to prevent flicker.

Resolution is an important indicator of the image quality in optical imaging systems. High resolution enables the system to capture detailed images, which is crucial for OET application experiments such as the biological imaging or observation of microscopic substances. However, in multi-channel or multi-light source systems, the consistency of the focal plane (that is, the focus of all image sensors is on the same plane) is also a problem that cannot be ignored. If the focal planes of different sensors are inconsistent, it may cause image blur and reduce imaging quality.

In order to ensure high-resolution images, it is necessary to first select suitable optical components to ensure that they have sufficient optical performance and resolution capabilities (such as high-quality lenses and sensors). In a multi-light source or multi-sensor system, in order to maintain the consistency of the focal plane, a precise focusing system or optical element correction method is usually used to ensure that the focus of all sensors is on the same plane. The optical elements can be precisely adjusted by the laser alignment technology, or the focal plane position can be adjusted in real-time during the experiment by using the automated focusing technology, thereby maintaining the overall imaging quality of the system.

The OET chip is also an important component of the experimental device, and the consistency of the chip has become an important challenge in the OET experiment. The performance of the OET chip directly determines the overall stability and accuracy of the OET manipulation experiment. In the manufacturing process of the OET chip, micron-level or even nanometer-level manufacturing errors will affect the performance of the chip. Different batches of chips may have slight structural differences, resulting in the inconsistent performance of each chip in the same optical system, which in turn affects the stability of the entire experiment.

In order to overcome this problem, strict quality control measures are usually adopted in the manufacture of OET chips, including high-precision lithography technology, sputtering coating technology, etching deposition technology, and precision measurement tools to ensure that the manufacturing process of each chip is consistent. In addition, the batch differences of chips can be adjusted through the post-calibration and compensation technology.

During the experiment, the OET chip will generate heat due to power supply and light irradiation, causing the temperature of the chip to fluctuate. If the chip works in different temperature environments, it may cause performance differences and affect the experimental stability of the entire system. Therefore, a precise temperature control system (such as thermoelectric cooling elements or temperature-controlled water cooling systems) is usually used to monitor and adjust the chip temperature in real-time. In addition, the design should also consider the heat diffusion design to ensure that the heat of the chip can be evenly distributed when working, avoiding inconsistent performance caused by excessive local temperature.

5.4.2 Software System

In the OET experiment, the software system is not only responsible for data acquisition and analysis but also plays a vital role in experimental control, light pattern projection, and result visualization. With the increasing complexity of the experiment, especially under the demand for optical path chips, image acquisition, and high-throughput, large-volume operations, the software system faces many challenges. These challenges are mainly reflected in the real-time control and modularization of multiple interfaces, and big data processing and storage problems are also serious.

In the OET experiment, image acquisition and light pattern projection are one of the core functional modules. The image acquisition interface is responsible for transmitting the real-time image data from the imaging device to the computer system for processing and analysis, while the light pattern projection interface is responsible for controlling the projection system to accurately project the light pattern onto the experimental sample. These two interfaces need to be able to process large amounts of data quickly and accurately to ensure the real-time and accuracy of the data.

However, image acquisition and light pattern projection usually require high frame rates and low latency, especially in high-speed imaging and dynamic experiments, where any delay may affect the accuracy of the experimental results. Different image acquisition devices and projection instruments may use different communication protocols and data formats, making interface compatibility and data transmission efficiency a challenge. Image acquisition and light pattern projection need to maintain strict synchronization to ensure the precise coordination of imaging and projection; otherwise, it may cause positional deviation of the projection pattern.

In order to meet these challenges, researchers usually use efficient data transmission protocols and hardware interfaces (such as USB3.0, GigE Vision, and Camera Link), which can provide high-bandwidth, low-latency transmission performance to meet the needs of high-speed image acquisition and real-time projection. At the same time, the hardware synchronization technology is used to ensure precise synchronization between image acquisition and light pattern projection through dedicated synchronization signal lines or synchronization software protocols (such as synchronization triggering and timestamp marking). When designing interfaces, choosing open standards and common protocols, such as GigE Vision or GenICam, can improve the compatibility and scalability of the system and reduce the difficulty of adapting between devices.

The interface of the automated control program is a key link in controlling hardware devices (such as optical path adjustment, chip operation, and projection control) in experiments. The real-time regulation of experimental equipment requires the software system to respond to the hardware in a short time and perform precise operations and feedback.

The automated control program must ensure real-time response and accurately control the operation of the hardware within microseconds, which is crucial for dynamic experiments (such as laser scanning and beam adjustment). Experiments usually involve the simultaneous operation of multiple hardware devices, such as the simultaneous control of image acquisition, light pattern projection, and chip adjustment. How to ensure the coordination of these hardware and avoid conflicts is a challenge. The hardware devices in the experiment may fail or be abnormal, and the control program needs to be able to respond and handle these problems in a timely manner to ensure that the experiment is not affected.

To solve the problems of real-time and accuracy, real-time operating systems (RTOSs) or multi-threaded programming methods are usually used to handle high-priority hardware control tasks. Through real-time scheduling strategies, ensure that each control task can respond and execute in a timely manner. In terms of concurrent control, a centralized or distributed control architecture can be adopted to achieve the coordinated work of multiple devices through standardized interfaces (such as Modbus, OPC, and EtherCAT) to ensure that hardware operations will not conflict. For hardware failures and abnormal feedback, a perfect error detection mechanism and automatic recovery program are designed. Once a hardware failure occurs, it can quickly capture and give feedback to prevent experimental interruption.

With the increasing the complexity of experimental systems, the modular design with multiple interfaces has become a common architecture for OET software systems. Each module is usually responsible for different functions, such as data acquisition, optical path control, result display, and data analysis. These modules need to communicate through standardized interfaces.

Different modules may be implemented by different software or hardware platforms. How to ensure that the interfaces between these modules are standardized and communicate effectively is a challenge. When multiple devices or systems need to be integrated in the experiment, there may be protocol incompatibility or interface mismatch between different hardware and software. As the experimental requirements change, the modular design of the software system must support the expansion and upgrade of the system to avoid the inability of the entire system to operate normally due to module incompatibility.

Therefore, it is necessary to adopt a modular design approach, and each functional module is independently developed and uses standardized interfaces (such as WebSocket and MQTT) for communication. This enables flexible collaboration between different modules, while reducing the dependency and coupling between modules, and improving the maintainability and scalability of the system. By using cross-platform development tools (such as MATLAB, LabVIEW, and Python) and general hardware interfaces (such as PCIe, USB, and Ethernet), compatibility issues between different devices can be resolved, ensuring the cross-platform support and hardware compatibility of the system.

With the rapid increase in the amount of experimental data, how to efficiently process and store massive data has become a major challenge in the experimental system. The data generated by the OET experiment usually includes high-resolution images, time series data, and a large number of experimental parameters, and traditional data processing and storage methods can no longer meet the needs.

A large amount of data will be generated in the experiment, especially in the case of high-resolution imaging and long-term tracking, the amount of data is very large. Traditional local storage methods are difficult to cope with the rapid growth of data, which can easily lead to insufficient storage space or storage access bottlenecks. For the processing of massive data, traditional computing methods may result in too slow processing speed to meet the needs of real-time analysis and decision-making.

In order to solve the problem of data processing and storage, many laboratories have adopted the cloud computing technology and distributed the data processing architecture. By uploading data on to the cloud platform, not only can the data be stored efficiently but also cross-platform sharing and remote access can be achieved. At the same time, the use of efficient data compression and indexing technology can effectively reduce storage space and increase the speed of data query. For data processing, the use of the parallel computing technology and high-performance computing clusters can greatly improve the efficiency of data processing and meet the needs of real-time analysis in experiments.

The software system plays a vital role in the OET experiment, especially in data acquisition, control system, and result visualization. Faced with the increasing complexity of experimental equipment, the rapid growth of data volume and the demand for high-throughput and large-scale operations, the software system must solve many challenges such as interface design, real-time control, system integration, and data processing and storage. Through the modular design, standardized interfaces, cloud computing architecture, and efficient data processing technology, these challenges can be effectively met and more stable, flexible and efficient support can be provided for experiments.

References

1 Zhang, S., Xu, B., Elsayed, M. et al. (2022). Optoelectronic tweezers: a versatile toolbox for nano-/micro-manipulation. *Chemical Society Reviews* 51 (22): 9203–9242.

2 Gan, C., Zhang, J., Chen, B. et al. (2024). Optoelectronic tweezers micro-well system for highly efficient single-cell trapping, dynamic sorting, and retrieval. *Small* 20 (23): e2307329.

3 Liang, S., Sun, J., Zhang, C. et al. (2022). Parallel Manipulation and flexible assembly of micro-spiral via optoelectronic tweezers. *Frontiers in Bioengineering and Biotechnology* 10: 868821.

4 Chiou, P.Y., Ohta, A.T., and Wu, M.C. (2005). Massively parallel manipulation of single cells and microparticles using optical images. *Nature* 436 (7049): 370–372.

5 Huang, K.-W., Su, T.-W., Ozcan, A., and Chiou, P.-Y. (2013). Optoelectronic tweezers integrated with lensfree holographic microscopy for wide-field interactive cell and particle manipulation on a chip. *Lab on a Chip* 13 (12): 2278–2284.

6 Hyundoo Hwang, H.C., Choo, J., and Park, J.-K. (2010). Optoelectrofluidic sandwich immunoassays for detection of human tumor marker using surface-enhanced Raman scattering. *Analytical Chemistry* 82 (18): 7603–7610.

7 Yang, W., Yu, H., Li, G. et al. (2017). High-throughput fabrication and modular assembly of 3D heterogeneous microscale tissues. *Small* 13 (5): 1602769.

8 Chu, P.-Y., Liao, C.-J., Hsieh, C.-H. et al. (2019). Utilization of optically induced dielectrophoresis in a microfluidic system for sorting and isolation of cells with varied degree of viability: demonstration of the sorting and isolation of drug-treated cancer cells with various degrees of anti-cancer drug resistance gene expression. *Sensors and Actuators B: Chemical* 283: 621–631.

9 Liang, S.Z., Cao, Y.Q., Dai, Y.G. et al. (2021). A versatile optoelectronic tweezer system for micro-objects manipulation: transportation, patterning, sorting, rotating and storage. *Micromachines* 12 (3): 16.

10 Liang, S.Z. (2020). *Based study on Microorganism Robot Control via Multifunctional Photoelectronic Tweezers System*. Beijing, China: Beihang University.

11 Gan, C.Y. (2021). *Research on Non-contact and Massively Parallel Manipulation Based on Optoelectronic Tweezers*. Beijing, China: Beihang University.

12 Cao, Y.Q. (2021). *Construction of Optoelectronic Tweezers System and Its Application in vitro Cell Sorting for the Achievement of High-throughput and High-precision*. Beijing, China: Beihang University.

13 Zhao, J.W. (2022). *High-throughput Single Cell Screening Based on Optoelectronic Tweezers*. Beijing, China: Beihang University.

14 Huang, S.X. (2024). *Basic Research on Sperm Screening Technology Based on Optoelectronic Tweezers System*. Beijing, China: Beihang University.

15 Lin Feng, "Miro Nano Robot", http://www.weinadongli.com/Default.aspx (accessed 25 January 2025).

6

Applications in Biology

Since OETs were invented, they have found extensive use in numerous biomedical applications due to their precision and versatility. This section will delve into how the OET technology has been utilized across various areas, including cell sorting, cell patterning, DNA transfection, and other significant medical procedures.

6.1 Manipulation of Biological Entities

6.1.1 Cell Sorting

One of the primary applications of OET is cell sorting, which has attracted significant research interest in recent years. Zhao et al. have developed a deep learning algorithm based on YOLOv5 for the high-throughput detection of single cells using an OET system [1]. The algorithm developed in this study allows for the rapid prediction of particle positions based on captured images. Following 361 training iterations, the mean average precision (mAP@0.5) of the precision-recall curve reached an impressive 98.1%. A single-cell separation platform was established by creating a 100-μm-thick micrograting structure using 3D printing technology. Experimental validation was carried out using the developed single-cell detection algorithm and simulation results. The system successfully achieved the automatic alignment of disordered microspheres or their distribution to the nearest micrograting within a short period. This capability demonstrates the potential of the OET system for high-throughput cell separation and the long-term dynamic observation of single cells. The application of this system is expected to significantly contribute to the study of cellular heterogeneity and the discovery of biopharmaceuticals.

OETs can also be utilized for cell manipulation to achieve direct contact between cells, which is highly valuable for studying the interaction between immune cells and other cell types, enabling the real-time analysis of the immune cell behavior. Ke et al. have developed an innovative cancer immunotherapy microenvironmental LabChip equipped with OET based on titanium oxide phthalocyanine (TiOPc), providing a stable and static microenvironment for the study of cell–cell interactions and real-time analysis of natural killer (NK) cell behavior [2]. As shown in Figure 6.1a-d, the utilization of the OET technology to bring NK

Optoelectronic Tweezers for Manipulation at the Micro and Nano Scale, First Edition.
Lin Feng, Menglu Tan, Ao Wang, and Jiaying Zhang.
© 2026 WILEY-VCH GmbH. All rights reserved, including rights for text and data mining and training of artificial intelligence technologies or similar technologies. Published 2026 by WILEY-VCH GmbH.

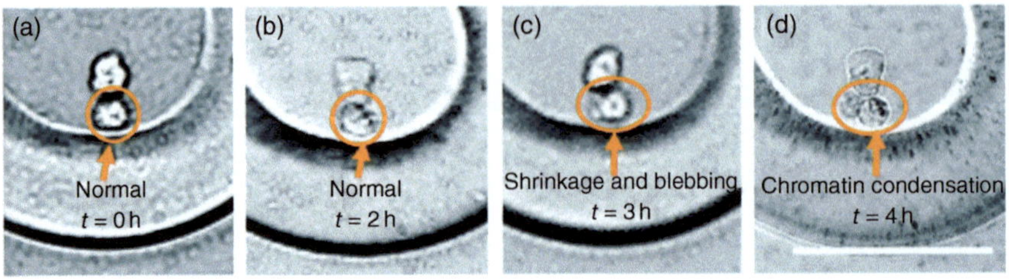

Figure 6.1 Time-lapse images of a target cell post-contact with a natural killer cell using OET. (a) Normal morphology (0 h); (b) normal morphology (2 h); (c) shrinkage and blebbing (3 h); (d) chromatin condensation (4 h). *Source*: [2] / The Royal Society of Chemistry.

cells (NK 92 MI cell line) into target cells (K562-EGFP cell line), allows for the investigation of the immunocytotoxicity of immune cells. When NK cells interact with target cells through OET, the target cells display apoptotic features such as cell shrinkage, bubbling, and eventual cell death within hours [2]. The study reveals that the percentage of NK cells effectively killing target cells using the OET method is nearly double compared to traditional detection methods. This enhanced efficacy is primarily attributed to the advantages of OET, which facilitate direct cell-to-cell interactions, in contrast to the random interactions that occur with traditional methods. The controlled and precise nature of OET enables a deeper understanding of the NK cell activity and cell interactions. OET can be further employed to manipulate different cell types and investigate their behavior under various experimental conditions. Additionally, cells have been utilized as test samples in other studies to evaluate and enhance the performance of different OET systems. These research endeavors play a critical role in advancing OETs as a versatile tool for cell manipulation and studying cellular interactions.

The ability to manipulate cells makes OET an important tool for applications in cell sorting and cell separation. Huang et al. reported a novel microfluidic-integrated OET platform for a single-cell sample preparation and analysis [3]. The platform described integrates OET with microfluidics by incorporating single-walled carbon nanotube (SWNT) electrodes into multilayer polydimethylsiloxane (PDMS) structures. This integrated system enables the selective manipulation and sorting of individual cells from a cell population based on their size, shape, and optical properties, such as fluorescence, using OET. Optically-induced dielectrophoretic (ODEP) forces are employed to transport the selected cells to isolation chambers within the microfluidic device. Figure 6.2a-d depict the sorting process of target cells from a cell mixture using this integrated platform, with the target cells being fluorescently labeled [3]. The sorting process involves several steps:

1) Visual inspection of cells under a fluorescence microscope to identify target cells based on their fluorescence.
2) Using optical traps to manipulate and move the target cells into a branching channel.
3) Flushing the main channel with air to generate nanoliter droplets.

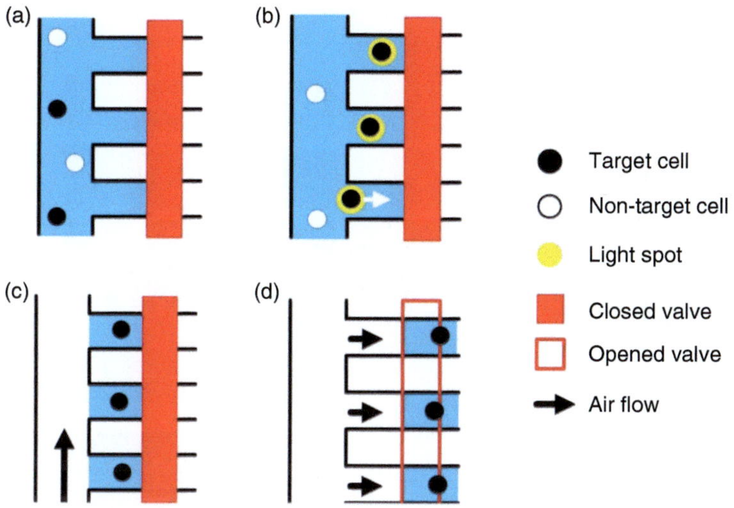

Figure 6.2 Diagram of the OET extraction of target cells in a microfluidic channel. (a) Inject cell mixture into the main channel; (b) move target cells to branch channels; (c) flush the main channel with airflow; (d) open the pneumatic valve. *Source*: [3] / Royal Society of Chemistry.

4) Opening a pneumatic valve to flush the main channel with air again, causing the airflow in the branching channel to generate air pressure that pushes the droplets to the outlet for efficient extraction.

Once target cells have been extracted, subsequent analytical methods like polymerase chain reaction (PCR) and DNA sequencing can be utilized to conduct additional analysis and cell characterization [3].

This sorting strategy involves the use of fluorescent dyes to selectively label specific cells in the mixture, enabling the identification of target cells before sorting them using OET. Immunofluorescently labeled circulating tumor cells (CTCs) have been effectively separated from blood samples, achieving complete purity, which allows for subsequent gene expression studies. Additionally, there are numerous methods available to differentiate cells by their inherent physical characteristics, including size and shape.

Chiu et al. [4] introduced a method using OET to enhance CTC isolation in a microfluidic system (Figure 6.3a–l). This approach aims to achieve high-purity CTC isolation beyond the capabilities of existing techniques. In this study, an ODEP-based microfluidic system was developed to further purify CTCs following conventional isolation methods. The system utilizes inherent laminar flow in a microchannel to transport cell suspension samples while fluorescent imaging identifies and positions target cancer cells. ODEP forces are then employed to precisely separate cancer cells from leukocytes.

Liao et al. [5] proposed a microfluidic method utilizing OET for the high-purity isolation of CTCs that have undergone epithelial-to-mesenchymal transition (EMT), specifically focusing on the CD45neg/EpCAMneg cells, which are crucial

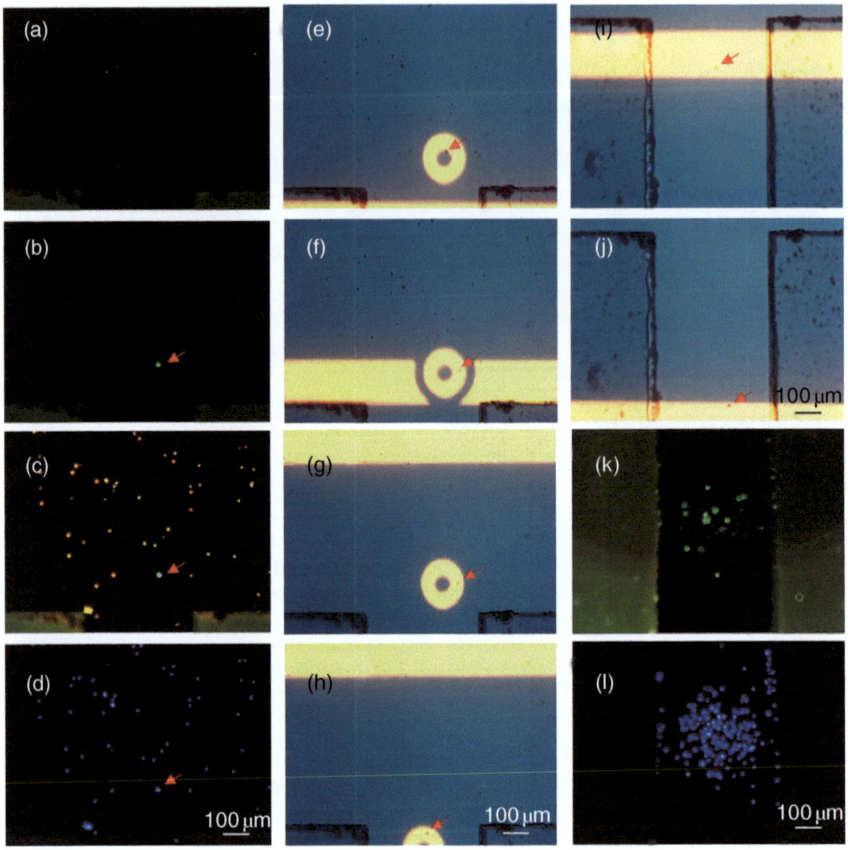

Figure 6.3 The overall CTC isolation and purification processes. (a, b) Use fluorescence to detect and pause flow for cancer cells (green dots); (c, d) identify leukocytes (red), cancer cells (green), and nucleated cells (blue); (e–g) separate cancer cells with circular light; sweep leukocytes aside; (h) move cancer cells to a side channel; (h–j) transport cancer cells near through-hole; (k, l) verify cancer cell purity using immunofluorescence. *Source*: [4] / Springer Nature / CC BY 4.0.

for studying cancer metastasis (Figure 6.4a–i). Traditional isolation techniques often miss these cells due to their downregulated EpCAM expression, but the integration of fluorescence microscopic imaging with ODEP forces within a microfluidic system offers a solution.

In this study, the microfluidic system leverages immunofluorescence dye staining and fluorescent microscopic imaging to recognize and locate target cells, followed by the ODEP manipulation for precise isolation. The system's design, including circular ring dark image patterns, prevents the unwanted adhesion of surrounding cells, ensuring contamination-free isolation. Experiments determined the optimal conditions for ODEP manipulation, with a 50% diameter ratio of target cell to circular light image and a 60-µm-width ring belt that effectively repels unwanted cells, such as leukocytes. The study demonstrates the capability of this approach to achieve 100% purity in isolating CD45neg/EpCAMneg cells from the blood samples of cancer patients, a level unattainable with current techniques.

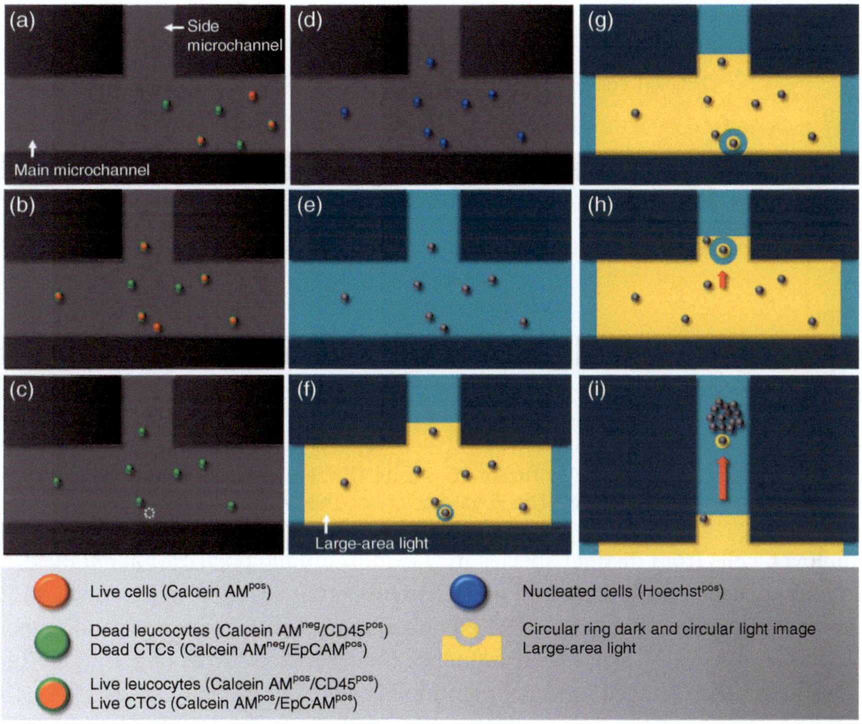

Figure 6.4 Schematic of the overall working procedures for the isolation of $EpCAM^{neg}$ and $CD45^{neg}$ cells. (a) Detect live $EpCAM^{neg}/CD45^{neg}$ cells (orange dots) with fluorescence; (b) pause flow when detected; (c, d) switch filters for green (leucocytes) and blue (nucleated cells) dots; (e) identify target cell with light microscopy; (f, g) use light to anchor and enclose target, expand ring; (h, i) transfer cell to side channel for collection. *Source*: [5] / MDPI / CC BY 4.0.

OET can also perform label-free sorting based on the size of different cells. Huang et al. proposed a microfluidic platform based on OET that is capable of separating CTCs in a continuous, label-free, cell-friendly, and high-purity manner [6]. As shown in Figure 6.5, the platform showcases the varying movement speed and manipulation force of cells across different species [6]. Prostate cancer cells (PC-3) and oral cancer cells (OEC-M1) are larger in size compared to white blood cells, resulting in stronger manipulation forces and increased movement speed. As a result, the OET system can effectively segregate cells based on their size.

Chou et al. [7] introduced a method using OET for the high-purity isolation of CTCs in a microfluidic system, enhancing the efficacy of conventional CTC isolation methods (Figure 6.6a-i). OET was employed for cell manipulation and flow velocity control, leveraging cell size differences between CTCs and leukocytes to achieve superior separation. The microfluidic system employs a novel four-cascade cell isolation approach using optical light-based virtual cell filters across four zones, allowing for high-resolution separation and improved cell purity. This design avoids the common issue of cell clogging found in membrane- or filter-based techniques by periodically removing trapped cells and regenerating fresh virtual filters.

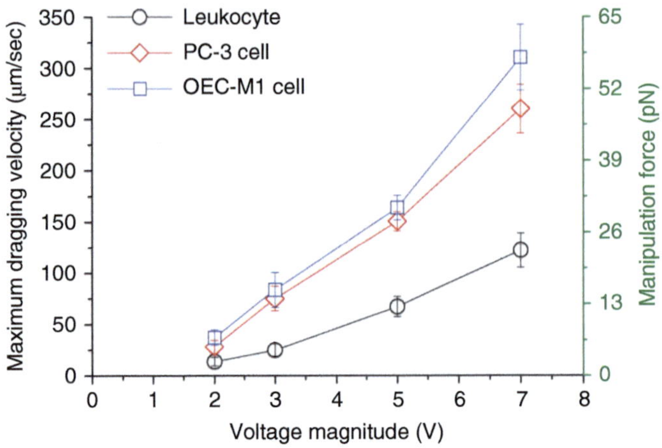

Figure 6.5 Measured moving velocities of PC-3 cells, OEC-M1 cells, and leukocyte cells at different bias voltages. *Source*: [6] / The Royal Society of Chemistry.

Moreover, cells from distinct species may exhibit unique polarization characteristics due to their morphology and dielectric properties (such as capacitance and conductivity), leading to different CM factors under specific frequency conditions. These variations in CM factors can influence manipulation forces (e.g. $F_{\mathrm{DEP}} \propto \mathrm{Re}[\mathrm{CM}]$) and can be leveraged for the separation of cells from different species [8].

OET can also be utilized for separating live cells from dead cells. Chiou et al. developed an ODEP technique that can create high-resolution electric field patterns on optoelectrically active surfaces [9]. This method allows for the specific enrichment of viable human B cells from a mixture containing both live and deceased cells, as depicted in Figure 6.7a-d. The distinction arises from the unique polarization traits of live and deceased cells, which stem from their differing membrane compositions. Viable cells possess functional cell membranes that are differentially permeable. Consequently, they can uphold varying ion concentrations within their intracellular and extracellular milieus through their membranes. Conversely, the membranes of deceased cells are unable to function as intended. Hence, when exposed to a culture medium with reduced ion levels, ions from within deceased cells will diffuse outward, resulting in the interior of these cells displaying electrical characteristics akin to those of an isotonic culture medium. This discrepancy in polarization traits between live and deceased cells emerges as a result of these processes. Consequently, in OET, live cells experience significantly greater manipulation forces compared to dead cells [9, 10]. Based on this principle, live cells can be effectively separated from dead cells (or abnormal cells with damaged cell membranes) using OET.

Hwang et al. [11] proposed a method using OET for discriminating normal oocytes in an optoelectrofluidic platform, leveraging ODEP forces for in vitro fertilization (IVF) applications (Figure 6.8). By integrating gravity with a pulling-up DEP force, induced by dynamic images projected from an LCD, the method effectively reduces friction forces acting on large and heavy oocytes, enhancing discrimination performance.

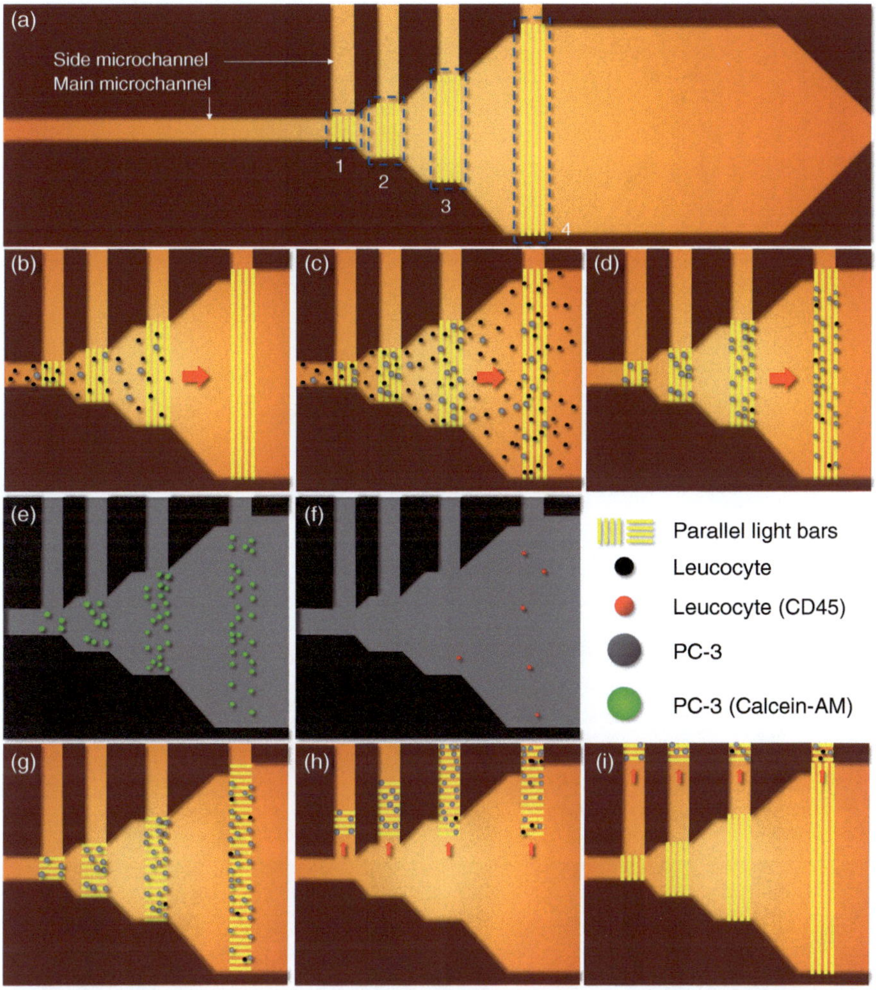

Figure 6.6 Layout and operational steps of an ODEP microfluidic system for isolating cancer cells using light bar arrays as virtual filters. (a) ODEP microfluidic system layout for cancer cell isolation with four light bar arrays as virtual filters; (b-i) cancer cell isolation process. *Source*: [7] / with permission of Elsevier.

The system applied a voltage condition of 10 V bias at 1 MHz to facilitate the movement of normal oocytes. By increasing the difference in moving velocity between normal and starved abnormal oocytes, the approach enables spontaneous discrimination of fertilizable oocytes under a moving image pattern. This gravity-compensated DEP strategy facilitates automatic and interactive selection, potentially improving IVF success rates by avoiding external contacts that could compromise the fertilization success.

Ohta et al. manipulated human sperm using OET, reviewing the electrophysical basis of the OET response to live and dead sperm, and measured the velocities of live and dead sperm, as shown in Figure 6.9, with live sperm exhibiting higher

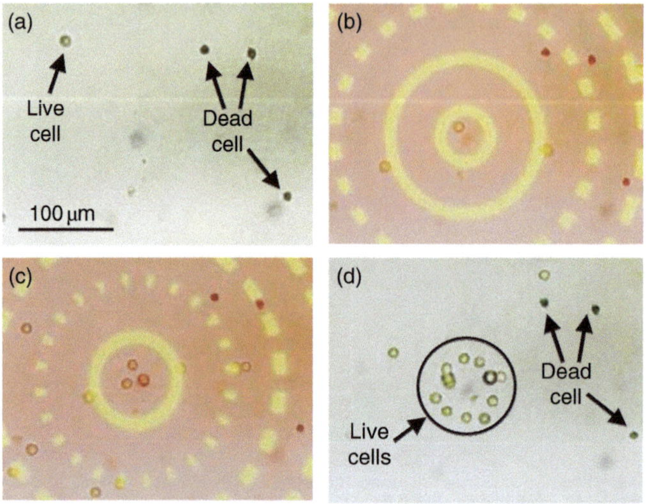

Figure 6.7 Isolation of live cells from a combination of live and dead cells. (a) Randomly positioned cells before OET; (b)-(c) cell sorting; (d) sorted cells. *Source*: [9] / Springer Nature.

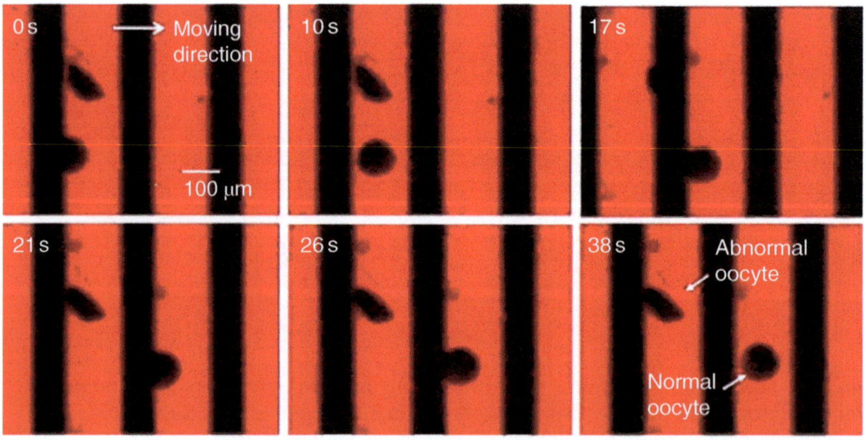

Figure 6.8 Discrimination of normal and abnormal oocytes using optical line scanning and antigravity manipulation. *Source*: [11] / AIP Publishing.

velocities than dead sperm [12]. In this study, OET was employed for sperm sorting and predicting sperm viability based on their dynamic responses [12]. Furthermore, OET has the capability to sort and separate drug-treated cancer cells with varying levels of drug resistance genes [13, 14]. Following the drug treatment, cancer cells with varying levels of drug resistance genes may undergo differing degrees of damage, impacting the integrity and permeability of their cell membranes. As a result, these cancer cells with varying levels of drug resistance genes demonstrate unique polarization characteristics post-treatment, leading to discrepancies in the strength of OET manipulation forces and movement velocities [13].

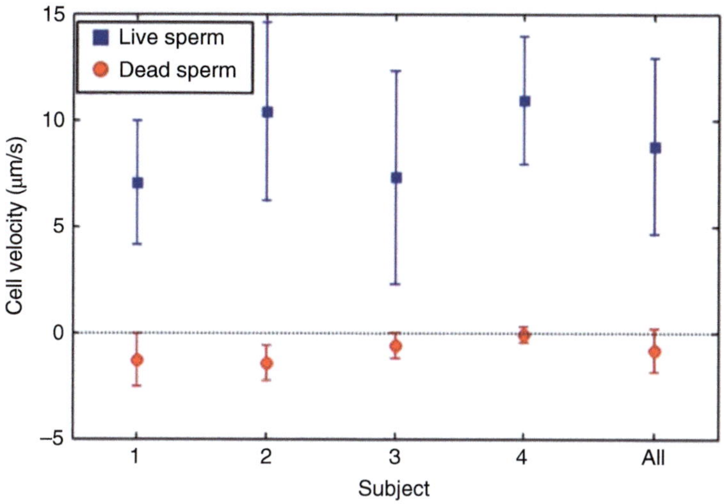

Figure 6.9 OET-induced velocities of live non-motile sperm and dead sperm. *Source*: Adapted from [6].

Chu et al. proposed a method that combines cell-based drug testing with ODEP-based cell manipulation. They designed and fabricated an OET microfluidic system for sorting, isolating, and separating drug-treated cells with varying levels of cellular viability [13]. Figure 6.10 illustrates the comparison of Dx5 cells (i.e. cells with drug resistance) and MES-SA cells (i.e. cells without drug resistance) following treatment with doxorubicin [13]. Following the drug treatment, Dx5 cells demonstrate notably increased movement velocities in contrast to MES-SA cells,

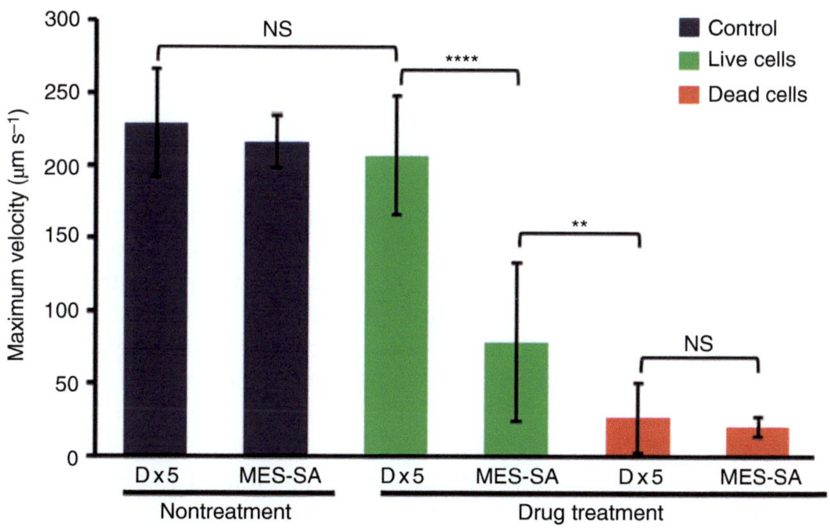

Figure 6.10 Maximum speeds of a moving light bar manipulating live and dead MES-SA and Dx5 cells pre- and post-treatment with 5.0 μg/mL doxorubicin for 48 hours. *Source*: [13] / with permission of Elsevier.

enabling the efficient segregation of these cell types. The findings suggest that OET can effectively differentiate cells according to their varied viabilities, attributed to differing levels of drug resistance within the cells. This sorting proficiency holds significance for both clinical practices and foundational scientific investigations. In addition to cells, OET can also be utilized for manipulating and sorting bacteria [15, 16].

Miccio et al. [15] introduced a method for manipulating and orienting *Escherichia coli* (*E. coli*) bacteria using OET. By leveraging the photorefractive (PR) properties of ferroelectric iron-doped lithium niobate crystals, this innovative strategy utilizes ODEP to immobilize and orient bacteria without the need for traditional electrodes, simplifying the process and enhancing versatility. The study demonstrated that light patterns could inscribe spatial charge distributions within the crystal, creating the electric fields capable of immobilizing approximately 80% of the *E. coli* bacteria on the crystal surface over an area of several square centimeters. Experiments confirmed successful bacterial trapping and orientation using this method, which exploits both direct laser irradiation and offline trapping to achieve high alignment percentages.

OET has been employed to detect antibiotic resistance in *E. coli* and to select smaller resistant bacterial clones within bacterial populations [16]. The operational principle resembles the sorting of cancer cells treated with drugs, as it involves using varying strengths of OET manipulation forces to differentiate bacteria with different levels of viability. The differences in bacterial survival are primarily attributed to different levels of antibiotic resistance [16]. Recently, OET has also been utilized for enriching and collecting membrane-enclosed extracellular vesicles released by cells, for disease diagnostics and drug delivery applications [17].

6.1.2 Cell Trapping

Cell trapping plays a crucial role in biomedical fields like regenerative medicine, drug screening, and tissue engineering. OET has proven to be a valuable tool for cell patterning [18], thanks to its capability to handle multiple cells at once. Gan et al. introduced an innovative electrical technique, designed as a microfluidic device, for arranging cells and selectively lysing target cells [19]. The microfluidic device is composed of a microcavity array situated on an optoelectric layer that employs the OET technique for simultaneous cell manipulation. As the cell suspension traverses through the microcavities, cells can be actively trapped within these structures through the influence of optically induced dielectrophoretic forces, creating a predetermined pattern array in under 120 seconds. Within the patterned cell array, the rate of single-cell capture exceeds 83%, with approximately 94% of the microcavities being occupied by cells. By adjusting the light patterns and applied signals, the optically induced DEP forces for cell capture can be experimentally modulated. Moreover, specific microcavities can be illuminated to electroporate and lyse targeted cells within a brief 5-second timeframe. The microcavity barriers and DEP forces serve to counteract the effects of flow, while the relatively enclosed space plays a critical role in preserving cell lysates. This innovative device, founded on OET and dynamic electroporation, facilitates the swift capture and precise lysis of cells at the individual-cell level, thereby advancing single-cell-based studies such as molecular diagnostics and disease detection.

Yang et al. developed a light-driven OET system utilizing titanium oxide phthalocyanine, an organic optoelectronic material, operating via negative dielectrophoresis (nDEP). With this innovative light-driven optoelectronic nDEP chip, they demonstrated the ability to pattern human liver cancer cells (HepG2) in a sucrose buffer solution [20]. By employing programmable light patterns, thousands of HepG2 cells were efficiently arranged into snowflake shapes within seconds. Comparable patterning results have been observed with other cell types in biocompatible hydrogel materials [21].

Lin et al. reported on the manipulation and assembly of alginate microbeads encapsulating cells using ODEP forces within a microfluidic perfusion cell culture system, designed for bottom-up tissue engineering applications [22]. Figure 6.11a-c presents a microscopic image of hydrogel microbeads containing chondrocytes at varied cell densities [22].

In this investigation, the chondrocytes of varying cell densities were enclosed within alginate microbeads using a microfluidic oscillator that operates on microfluidic technology. These encapsulated chondrocytes were then organized through the OET assembly to form a condensed lamellar cell culture structure. Subsequently, several layers of lamellar cell cultures were stacked to form larger structures that mimic the three-dimensional cell distribution similar to that of a natural articular cartilage. In addition to the direct assembly of hydrogel microbeads loaded with cells, a more accessible cell patterning method has been developed. This technique utilizes OET in conjunction with UV-curable poly (ethylene glycol) diacrylate (PEGDA) hydrogels [23]. In this study, OET was used to organize cell patterns within PEGDA hydrogels, which were then solidified through UV polymerization. The result of the UV polymerization process was the formation of

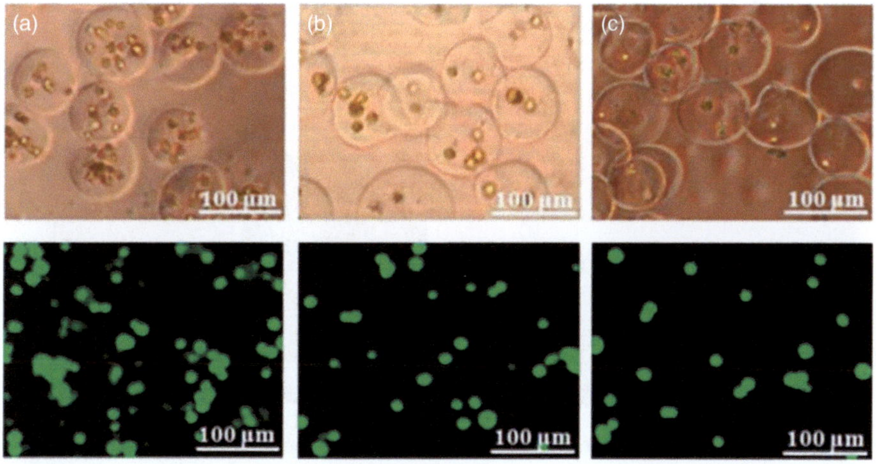

Figure 6.11 Bright-field and fluorescence microscopy images showing hydrogel microbeads encapsulating chondrocytes at varying cell densities. (a) 2.4×10^7; (b) 1.0×10^7; (c): 7.7×10^6 cells/mL. *Source*: [22] / The Royal Society of Chemistry.

hydrogel sheets encapsulating cell patterns, suitable for applications in drug screening and tissue engineering.

Three-dimensional hydrogel microstructure-containing cells have found extensive applications in microscale tissue engineering, personalized drug screening, and regenerative medicine. Nonetheless, the adoption of microtissue structures encounters challenges due to the absence of technology that integrates high-throughput manufacturing and assembly. In response to this, Yang et al. introduced a multifunctional methodology for producing personalized microstructures and reassembling components comprising functional units into unified geometric shapes [24]. Custom microstructures are created dynamically within microfluidic devices and then moved to a chip utilizing optically induced electrokinetics (OEK) for further manipulation and assembly. Moreover, components containing distinct cell types can be organized into specified geometric shapes with precise dimensions, offering potential applications in microscale tissue engineering. As depicted in Figure 6.12a, a schematic representation of a high-throughput microfluidic system illustrates the integration of efficient manufacturing and assembly of hydrogel microstructures. In this setup, UV-curable PEGDA microstructures are fabricated

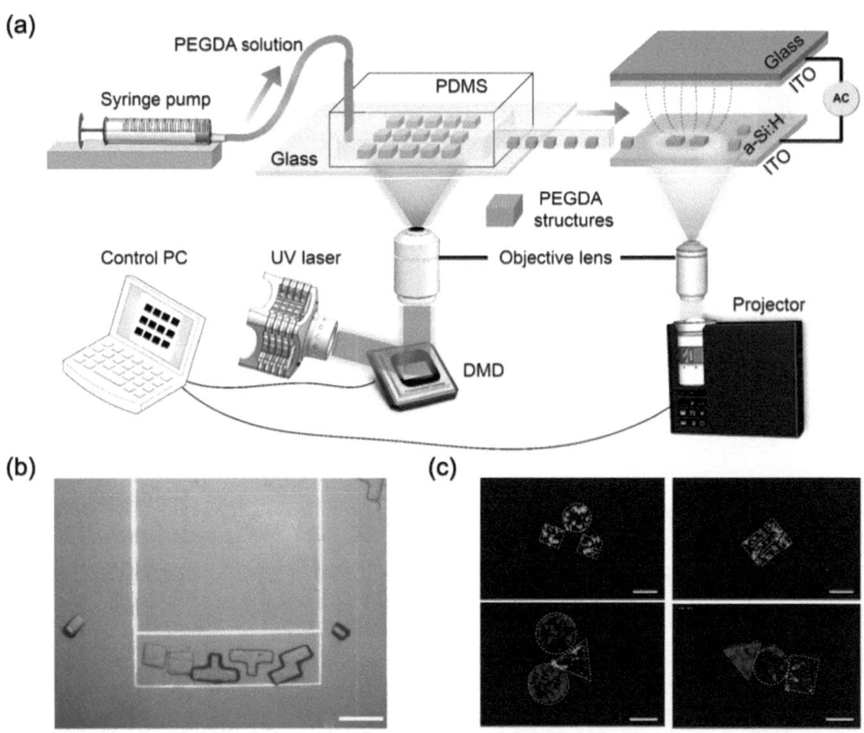

Figure 6.12 (a) Schematic diagram of a high-throughput manufacturing and flexible manipulation system, consisting of two main units: a hydrogel manufacturing unit based on DMD and a manipulation unit based on OET. (b) Microscopic images showing the construction of a production line with various PEDGA hydrogel microstructures using OET. (c) Assembly of cells encapsulated in PEDGA hydrogel microstructures. *Source*: [24] / John Wiley & Sons.

utilizing the maskless lithography technology, which relies on a DMD projector and UV laser [24]. The DMD serves as a reconfigurable mask, allowing for the dynamic control of UV light to create various shapes of PEDGA microstructures [24–26]. Following fabrication, the polymerized PEGDA microstructures are moved to an OET device for manipulation and assembly, as illustrated in Figure 6.12b. Additionally, cell-encapsulated PEGDA microstructures are created in precise shapes while maintaining cell viability and proliferation. These encapsulated cell microgels are structured in shapes that do not compromise cell viability and proliferation. Subsequently, these cell-encapsulated microgels are assembled into desired geometries utilizing OET, as shown in Figure 6.12c [24]. Thus, integrating UV/hydrogel-based fabrication with the OET-based assembly shows considerable promise for developing functional biostructures.

Beyond the use of UV light for hydrogel polymerization, OET has also proven effective for the electropolymerization and structuring of hydrogels without UV sources [27–30]. This technique capitalizes on the photoelectric effect within OET systems, allowing the electrochemical initiation of PEGDA hydrogel polymerization [28, 29]. In essence, electrons originating from the OET's photoelectric conductive layer interact with hydrogen ions in the PEGDA hydrogel, reducing these ions to hydrogen radicals. These radicals subsequently trigger a chain reaction, prompting PEGDA polymerization and hydrogel crosslinking. This method enables the fabrication of PEGDA hydrogel microstructures with precisely controlled dimensions and geometries, using non-UV optical images in the OET setup. A variety of light patterns can produce corresponding hydrogel designs, such as the parallel lines of differing widths, triangles, quadrilaterals, pentagons, circles, and five-pointed stars, as demonstrated in Figure 6.13a. For example, a bright star-shaped light pattern is used to create matching hydrogel designs, where the dark star shapes reveal

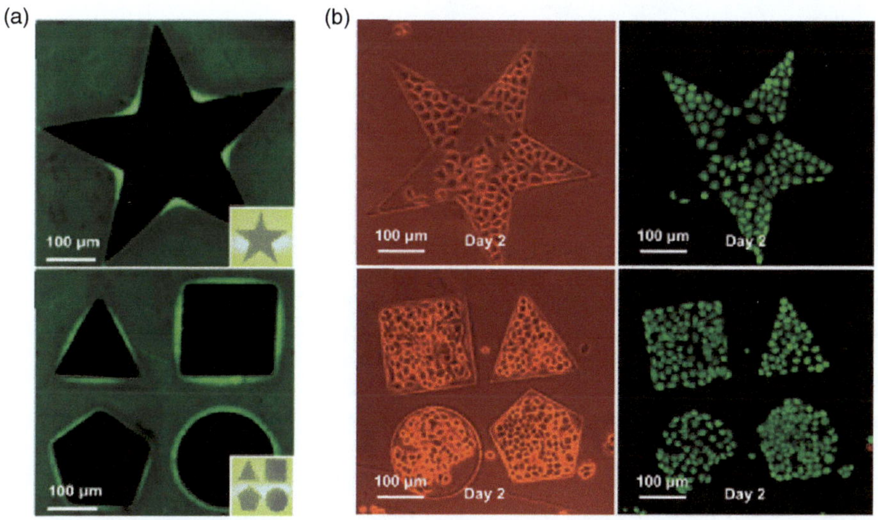

Figure 6.13 (a) Fluorescent images of different PEGDA structures, and pictures (insets) of the corresponding projected light patterns; (b) cell staining result. *Source*: [27] / The Royal Society of Chemistry.

exposed a-Si:H surfaces, while the surrounding green areas consist of crosslinked PEGDA hydrogels [27]. Once PEGDA hydrogel patterns are produced, they serve as physical scaffolds for arranging cells. As shown in Figure 6.13b, MCF-7 cells adhere exclusively to a-Si:H surfaces, avoiding the hydrogel regions, which results in the formation of varied cell patterns like five-pointed stars, triangles, quadrilaterals, pentagons, and circles. The selective adherence of cells is due to the innate non-adhesive properties of the PEGDA hydrogel material [27]. Consequently, OET is adept at fabricating PEGDA hydrogel microstructures to pattern cells into predetermined shapes, enhancing versatility in tissue engineering and other biomedical applications.

Furthermore, by adjusting the light exposure time and light distribution in OET, complex 3D hydrogel microstructures with controllable thickness have been successfully fabricated in a layer-by-layer manner [29]. Employing a comparable method, a sequence of dynamically modified images is utilized to generate hydrogel scaffolds conducive to tissue growth, resulting in a 3D mesh-like hydrogel network with customizable pores and gaps. Beyond the creation of hydrogel microstructures, OET has been leveraged to swiftly assemble structured microcolumn arrays from thin polydimethylsiloxane (PDMS) polymer films, as shown in Figure 6.14 [31]. This work exploits the electrofluidic instability effect, induced by photo-induced transverse electric fields and thermocapillary flow. As a result, OET can be used to transform PDMS thin films into periodic 3D column arrays and control the voltage for the column diameter and light patterns for column positioning [31, 32], providing a novel approach for the efficient and straightforward manufacturing of polymer microstructures for various biological micro-electromechanical system (MEMS) applications. Wang et al. [32] introduced a study on managing temperature

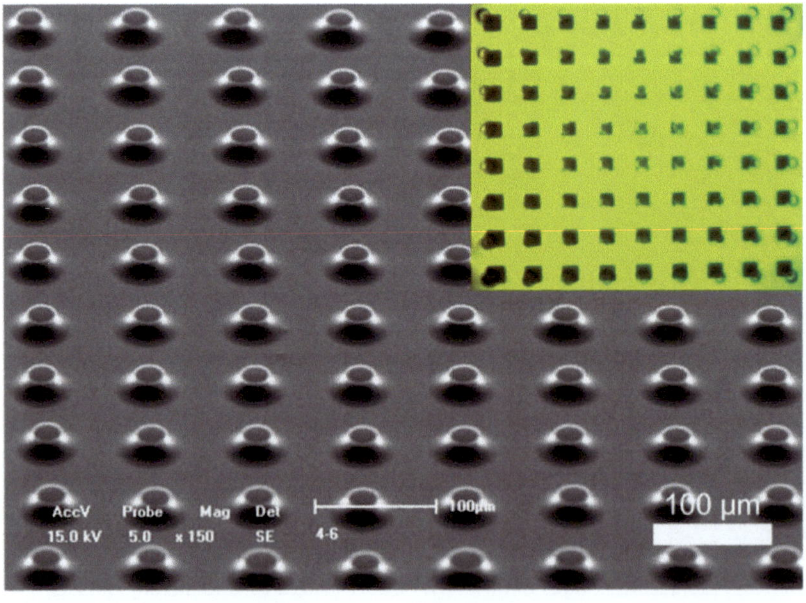

Figure 6.14 SEM images of the PDMS micro-pillar array. *Source:* [31] / AIP Publishing.

fluctuations in optically induced OEK-based technologies. The study systematically measures temperature changes and distributions in an OEK chip using an infrared camera, providing contactless, transient measurements. Key influencing factors such as light color, projected image shape and size, light distribution, and alternating current (AC) voltage parameters were examined. The findings indicate that the average temperature of a projected area is affected by these variables and highlights how the careful design of projected light patterns is crucial for controlling temperature distribution, especially in complex systems.

Moreover, Gan et al. revealed an innovative microfluidic platform utilizing OET to arrange and sort individual cells at a flow rate of 20 μL/min (Figure 6.15a-g) [33]. In the system, cells were individually confined within microcavities, achieving a high single-cell capture rate exceeding 91.9%. Through a detailed examination of factors such as illumination patterns, flow rate, and applied voltage, the single-cell capture rate was further improved. The integration of the OET system with a microcavity array showcased the adaptability and versatility of the device,

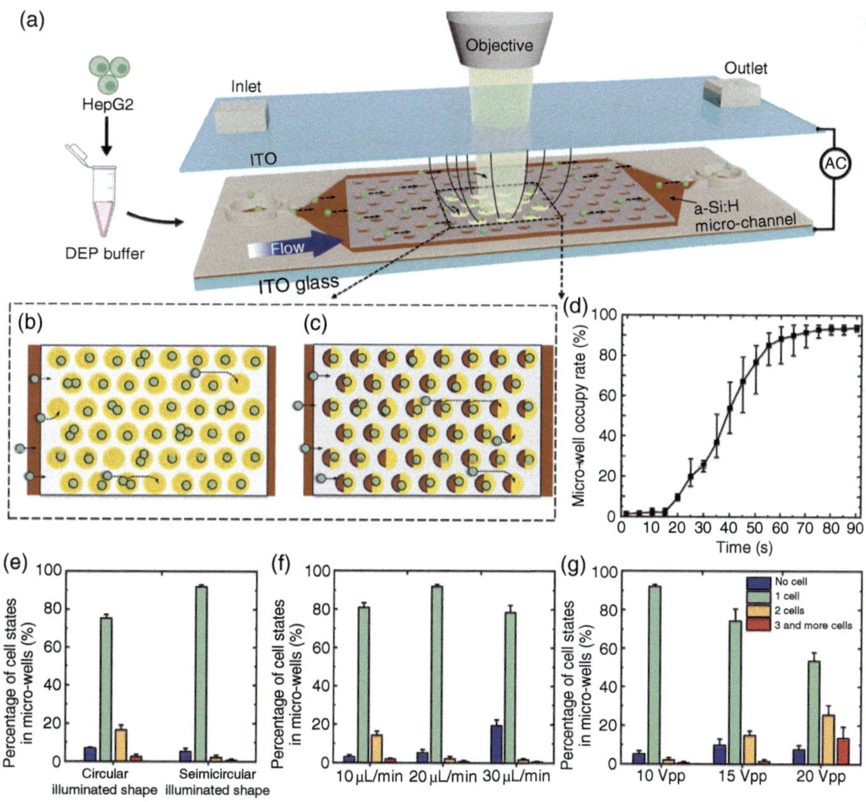

Figure 6.15 Schematic of the OET chip and cell occupancy based on illuminated shapes, flow rates, and DEP signal amplitudes. (a) A conceptual overview of the experimental setup; (b) circular light projection diagram; (c) semicircular light projection diagram; (d) micro-well occupancy rate over time; (e) cell occupancy with semicircular vs. circular shapes; (f) cell occupancy at different flow rates; (g) cell occupancy at various signal amplitudes. *Source:* [33] / John Wiley & Sons.

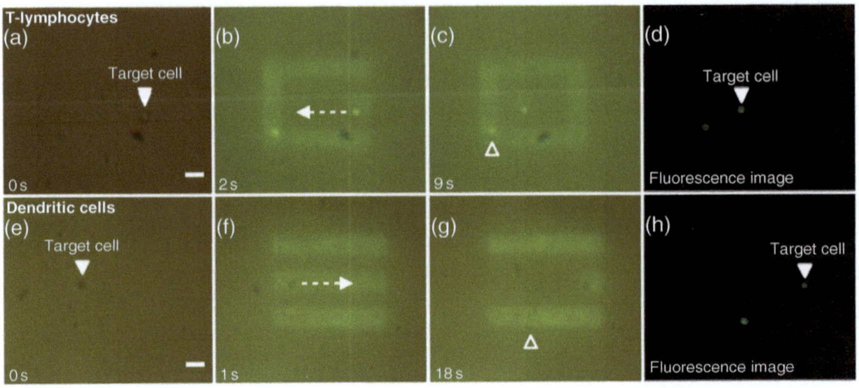

Figure 6.16 Achieving trapping and imaging of individual T-lymphocytes and dendritic cells. (a) and (e) Before micro-LED pattern turn on; (b) and (f) immediately after the start of a pixel animation; (c) and (g) after pixel animation; (d) and (h) fluorescence image showing CFSE labelled cells after pixel animation. *Source*: [34] / Optical Society of America.

enabling the simultaneous capture and isolation of specific cells based on their dielectric properties. Experimental findings demonstrated a sorting accuracy of over 91% for HepG2 and L-O2 cells. Furthermore, fluorescent markers were utilized to identify isolated, fluorescently labeled HepG2 cells and selectively release them within the chip. The platform's capability to efficiently capture and release individual cells significantly boosts its potential for future applications in biological research.

Jeorrett et al. introduced a miniaturized OET system for cell manipulation and analysis [34]. A complementary metal-oxide-semiconductor (CMOS)-controlled gallium nitride micro-LED array was employed to concurrently control the dielectric traps within the OET device and conduct the fluorescence imaging of contrast dye-labeled cells. This functionality allows for the direct identification, selection, and manipulation of interactions between individual T lymphocytes and dendritic cells. Both T lymphocytes and dendritic cells experience positive DEP in the OET device, with the target cells entering the optical cage along the path of a single illuminated pixel, while the optical cage walls prevent other cells from interfering with the target cell (Figure 6.16a–h). Subsequently, cell identification is accomplished through fluorescence imaging. These instances showcase the potential for intricate single-cell manipulation and the formation of immune cell arrays, with the option of utilizing fluorescence-based detection to elucidate subsequent cellular functions.

6.2 Biomedical Applications and Advancements

6.2.1 Immunoreaction Analysis

In addition to cell manipulation, OET is also utilized as an immunoreaction platform for biological assays. Han et al. illustrated a schematic of an immunoreaction system based on OET in Figure 6.17 [35].

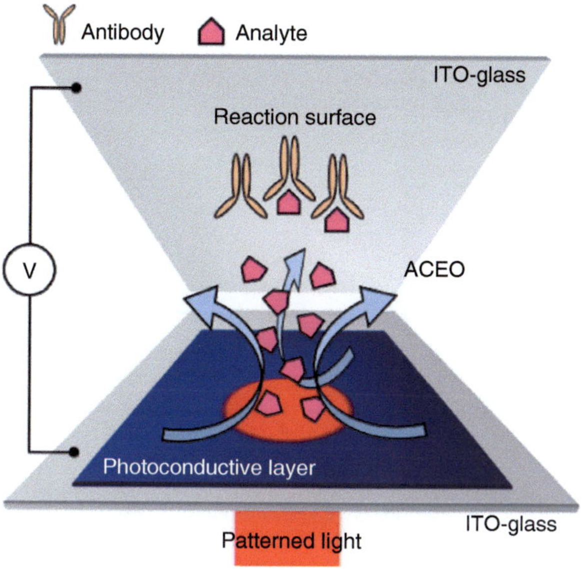

Figure 6.17 Schematic illustration of the optoelectrofluidic immunoreaction system. *Source*: [35] / Royal Society of Chemistry / CC BY-NC 3.0.

The system is constructed using two conventional ITO glass slides, forming a reaction chamber that holds approximately 5 μL of sample droplet. The top layer serves as the antibody-immobilized substrate, while the bottom layer functions as the photoconductive layer of the optoelectrofluidic device. When the AC voltage is applied, an illuminated light pattern on the photoconductive layer induces strong counter-rotating vortices, transporting analytes from the bulk solution to the vicinity of the assay spot on the glass substrate. This configuration addresses the slow immunoreaction issue faced by diffusion-based sensing systems, enhancing binding efficiency through the optoelectrofluidic method. They examined the impact of optically induced dynamic AC electroosmotic flow on optoelectrofluidic enhancement for surface-based immunoreactions through mathematical simulations and real experiments using immunoglobulin G (IgG) and anti-IgG. Results show that dynamic light patterns provide better immunoreaction efficiency than static light patterns due to the effective mass transport of target analytes. Under a growing circular light pattern, the system achieves a 2.18-fold enhancement compared to passive modes, as shown in Figure 6.18 [35].

Furthermore, it has been demonstrated that multiple detection points can be created in a microarray configuration on the OET top plate. The light-induced AC electroosmotic flow produced by the microarray-patterned light can efficiently transport analytes to these detection points simultaneously, thereby enhancing the overall assay efficiency [36]. Through the utilization of this setup, the immunoreaction time was decreased from over 30 minutes to just 10 minutes. Significantly, the successful detection of multiple proteins utilizing different IgGs was showcased, highlighting the capability to conduct multiple assays concurrently on the OET immunoreaction platform integrated with the microarray [36].

In addition to surface-based immunoassays, OET can also be utilized for immunoassays based on antibody-conjugated microbeads [37, 38]. Hwang et al.

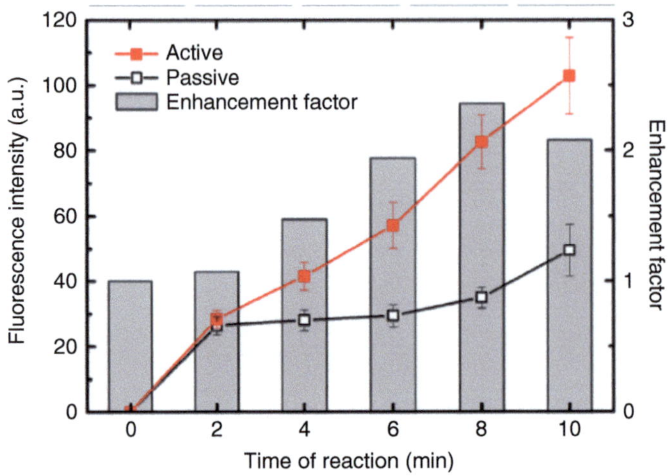

Figure 6.18 Fluorescence immunoreaction results of active and passive modes. *Source*: [35] / Royal Society of Chemistry / CC BY-NC 3.0.

presented a schematic diagram of a bead-based immunoassay procedure in the OET system in Figure 6.19a-e [37]. The assay comprises five steps: (1) sample injection; (2) the global concentration of antibody-conjugated microbeads and sample mixing; (3) washing of free analytes and probe nanoparticles; (4) local concentration of immune complexes; and (5) surface-enhanced Raman scattering (SERS) detection of probe nanoparticles on the immune complexes. With the exception of step 1, the remaining steps are carried out using light patterns and AC voltages with varying frequencies. Different AC frequencies lead to distinct primary electrokinetic mechanisms in the OET system, facilitating the execution of various assay steps. Leveraging microbead immunoassays, the quantitative and automated detection of alpha-fetoprotein was performed in 500 nL sample droplets. The detection limit achieved was approximately 0.1 ng/mL, underscoring the OET platform's efficacy

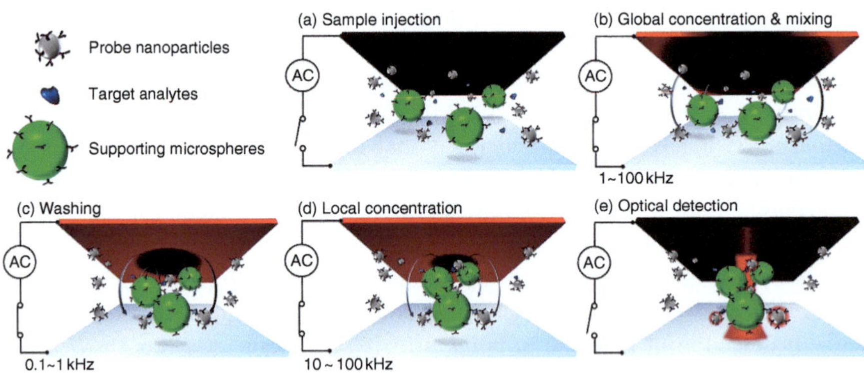

Figure 6.19 Schematic procedures of the immunoassay based on antibody-conjugated microbeads in an OET system. (a) Sample injection; (b) global concentration and mixing; (c) washing; (d) local concentration; (e) optical detection. *Source*: [37] / with permission of American Chemical Society.

as a robust optoelectronic immunoassay tool capable of rapid and highly sensitive detection of biochemical substances within minute sample volumes [37].

6.2.2 Cell Analysis

OET can be utilized for cellular analysis based on the dynamic responses of cells under different experimental conditions. For example, many cells exhibit a unique spinning behavior in the OET system, which can serve as unlabeled biomarkers for the rapid identification and characterization of different cell types [39, 40]. While the mechanisms underlying cell spinning in OET remain controversial [40–43], it has been demonstrated that spinning is determined by the electric field (i.e., frequency and field strength) and the polarization characteristics of the cells. Liang et al. reported a novel method that utilizes self-rotating unlabeled biomarkers in a microfluidic platform based on OEK to rapidly identify different types of cells and drug-treated cancer cells [39]. Results demonstrate the spinning speeds of four cell types (MCF-7, HEK-293, L929, and C2C12 cells) in relation to the applied AC frequency under a consistent bias voltage of 10 V. The distinct spinning spectra displayed by these cell types, are valuable for cell identification. The differences in spinning spectra are primarily caused by the cells' intrinsic dielectric parameters (such as cell membrane/cytoplasm/nucleus capacitance and conductivity), leading to varying polarization characteristics among different cell species.

The behavior of MCF-7 cells under varying drug concentrations was examined in terms of their spinning dynamics [39]. It was found that the cell rotation speed declined with an increase in the drug levels. This reduction in rotational velocity is mainly due to the drug's effects on the cell membrane. Elevated drug concentrations tend to inflict more substantial damage on the cell membrane, altering the polarization properties of the cells. As a result, MCF-7 cells exposed to higher doses of the drug exhibit slower spinning speeds.

Liang et al. [40] presented a study on the translational and self-rotational behaviors of Raji cells, a type of B-cell lymphoma, within a nonuniform and nonrotational AC electric field on an optically activated microfluidics chip (Figure 6.20a-n). The research utilizes digitally projected light patterns as virtual electrodes, forming nonuniform electric fields that enable both the movement and self-rotation of Raji cells. Experiments reveal that the cells exhibit maximum self-rotational speed at a frequency of 60 kHz, showing a quadratic relationship with the applied AC voltage. Through this approach, the membrane capacitance of the cells is determined by analyzing the crossover frequency spectrum. A pattern of four orthogonal electrodes projected onto the chip was utilized to further study cell behavior, revealing that cells located at the center remain static, while those positioned off-center undergo self-rotation. The numerical simulations of electric field distributions corroborate the experimentally determined nonuniform and nonrotational nature of the fields, underscoring their role in inducing self-rotation.

In addition to measuring the rotational speed of cells, OET can also be utilized for measuring the mass and density of single cells [44, 45].

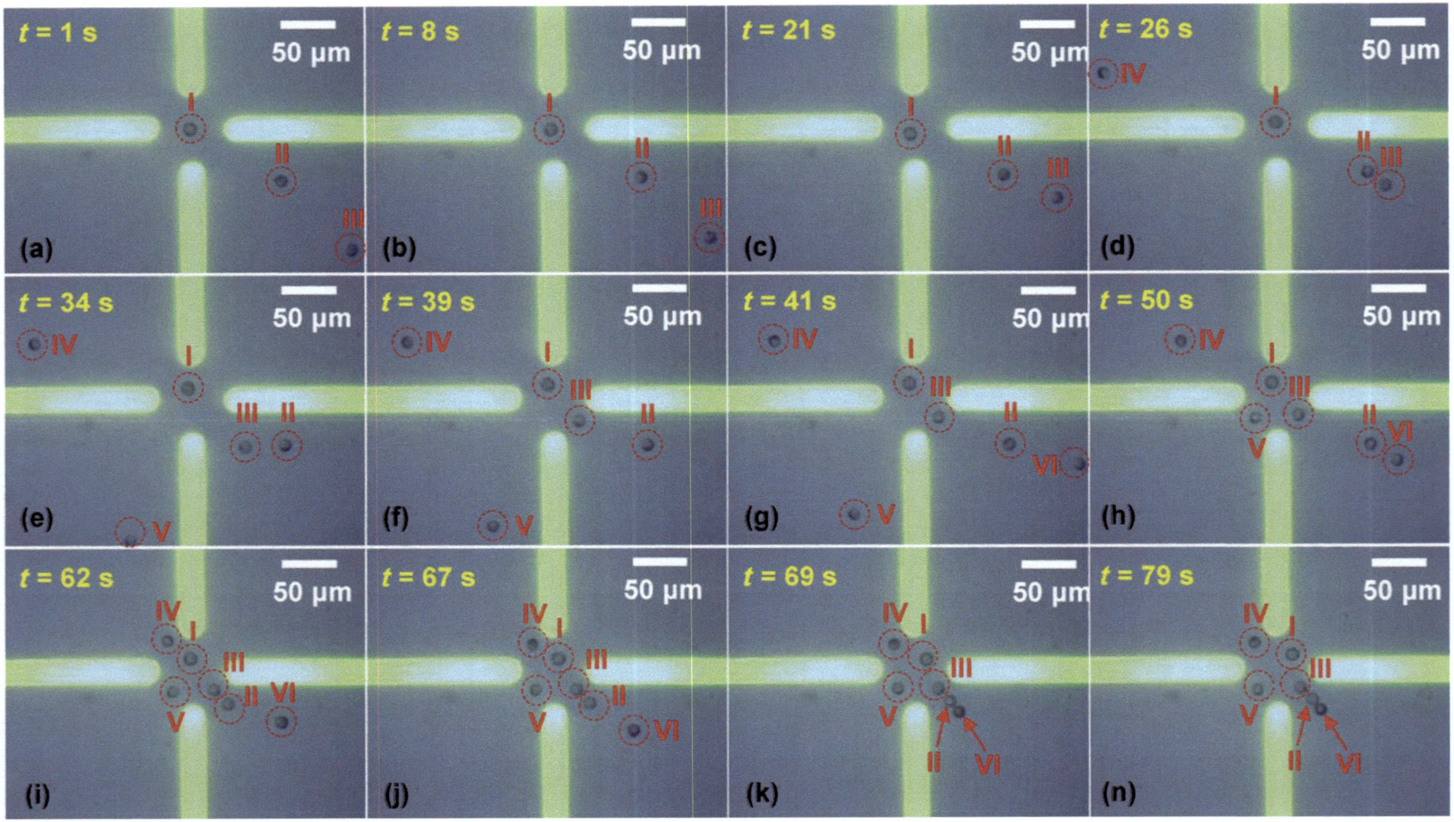

Figure 6.20 The self-rotation behaviors of Raji cells under optically projected four orthogonal electrodes at various applied frequencies and AC bias potentials. (a) Cells I–III visible, I is centered at 5 Vpp, 40 kHz; (b) cell I steady, III rotates; (c) at 65 kHz, I moves, II and III rotate; (d) at 70 kHz, I rotates and returns, IV moves centrally; (e) at 75 kHz, all move faster towards pattern; (f, g) cells III and VI move to center; (h–j) cells I, II, IV, V, VI rest in centers at higher frequencies; (k) cell VI at rest near cell II between electrodes; (n) cell I moves and rests at central area at 100 kHz. *Source*: [40] / AIP Publishing.

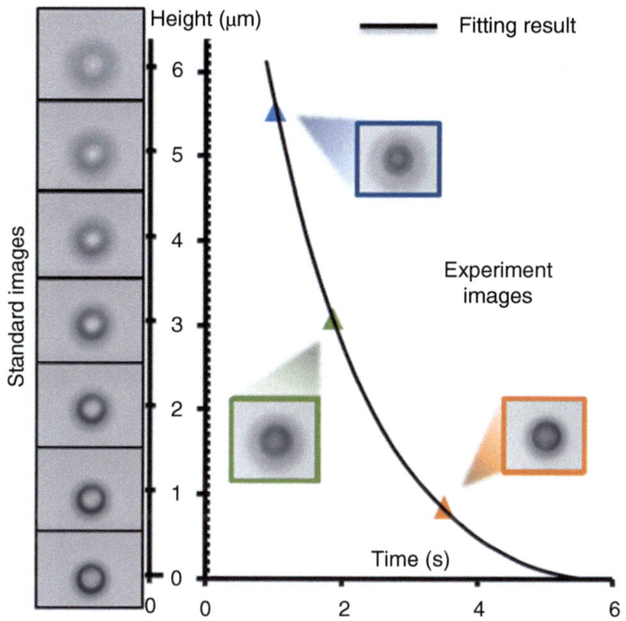

Figure 6.21 Illustration of particle height determination by matching the dynamic images with standard images of known heights. *Source*: [8] / Royal Society of Chemistry.

Zhao et al. reported a novel technique that utilizes the OEK microfluidic platform for the rapid measurement of the density and mass of single cells [44]. This technology is founded on the principles of the sedimentation theory, computer vision, and microparticle manipulation techniques within the OEK microfluidic platform. Initially, cells are vertically elevated to a specific height through OET manipulation forces. Subsequently, the cells are released to "free-fall" onto the OET cell substrate surface, where they contend with buoyancy, gravity, and Stokes drag forces. A computer vision algorithm is then utilized to swiftly ascertain the mass and density of each cell. This visual algorithm accurately monitors the cell's motion trajectory and leverages the sedimentation theory to compute the mass and density of the cells being measured. To verify the accuracy and efficacy of this approach, a theoretical model involving micro-sized spheres settling towards an infinite plane in a microfluidic environment was formulated and experimentally validated using standard micropolystyrene beads, as shown in Figure 6.21 [44]. Subsequently, the mass and density of yeast cells and leukemia cells were measured [44, 45], with results comparable to those obtained using the existing method of suspended microchannel resonator.

6.2.3 Cell Lysis

The OET technology serves as a pivotal instrument in the realm of cellular lysis, a process that is instrumental in the extraction of proteins and nucleic acids from within the cellular milieu. This technique is extensively harnessed across a multitude of scientific inquiries, thereby underscoring its significance in the advancement of biological and chemical research endeavors. Kremer et al. demonstrated an electrical method to selectively lyse live cells in different cell populations, where OET can selectively disrupt cells based on their morphology [46]. Figure 6.22a-d illustrates the brightfield microscopy images of small red blood cells (RBCs) and large white blood cells (WBCs) before, during, and after the activation of OET-induced light and electric fields. In this context, only RBCs undergo lysis, while the morphology of WBCs remains unchanged. This differential response is primarily attributed to the influence of the cellular shape, with RBCs being biconcave and WBCs being spherical, on the distribution of electric fields. The distinctive geometric profile of RBCs, which is broad and flat, leads to the significant deflection and distortion of electric field lines around the cell surface. As a result, the electric field becomes concentrated around and beneath the RBCs, creating hotspots that enhance the probability of electrolysis. This shape-dependent

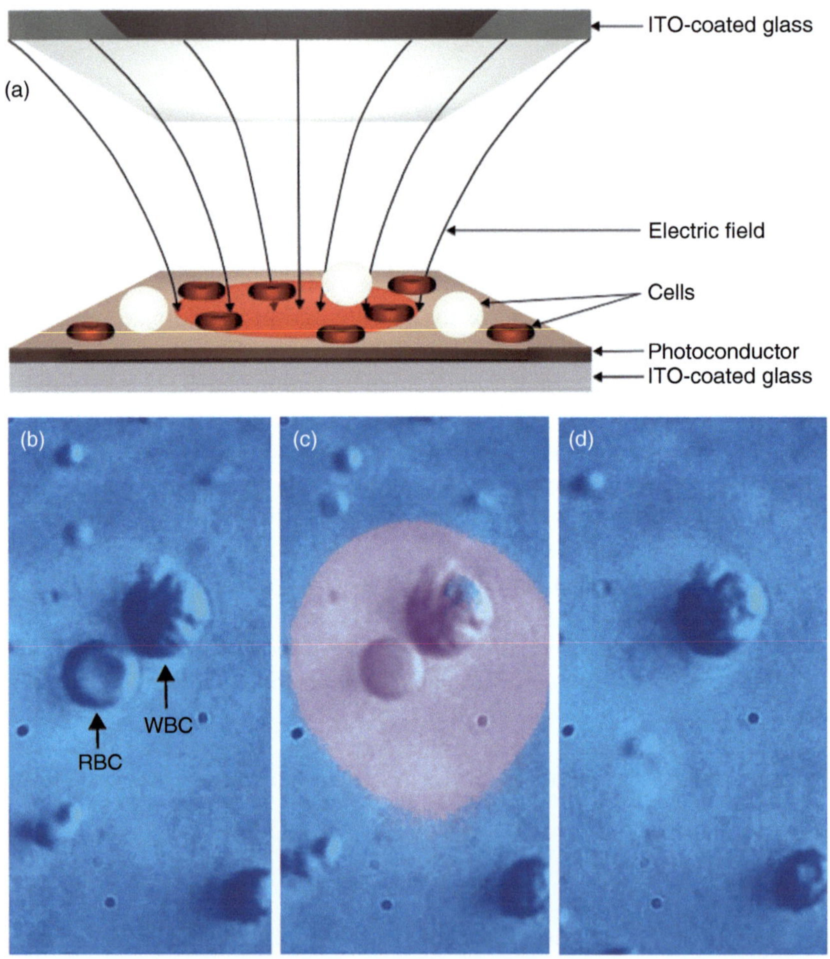

Figure 6.22 Shape-selective lysis using an optoelectronic system. (a) Diagram of the optoelectronic device with the electric field in the illuminated area; (b)–(d) smaller RBC and larger WBC before, during, and after light and field activation. *Source*: [46] / John Wiley & Sons.

lysis phenomenon is also observed when RBCs are co-mingled with parasites, underscoring the effectiveness of this technique [46]. Importantly, the OET technology enables the simultaneous processing of hundreds of cells over an extensive area, spanning several square millimeters.

6.2.4 Commercial Application

The inventor of OET, Wu Ming, established Berkeley Lights Inc. (BLI) [47]. This California-based company has successfully commercialized fully automated OET instrumentation for applications in antibody discovery [47–49], cell line development [47, 49–52], cell therapy development [47, 52, 53], and functional screening and selection of individual cells in synthetic biology.

The Optofluidic Platform developed by Berkeley Lights combines the nanoliter-scale microfluidic reactor technology (Nano-Pen) with opto-electropositioning (OEP), creating a versatile single-cell platform for sorting, analysis, culture, and export functions. The precursor to OEP is the OT technology, pioneered by Ashkin, a recipient of the 2018 Nobel Prize in Physics. OT uses a single laser beam focused through a high-magnification microscope objective to create an optical trap. Microscopic particles are trapped in the optical trap due to radiation pressure, and the manipulation of these particles is achieved at the piconewton scale. However, OT requires high-intensity laser light, which can potentially damage cells and limit the ability to work with multiple cells simultaneously.

To address these challenges, the optoelectronic positioning technology, OEP, was developed. OEP utilizes a light beam to generate an electric potential field on optically sensitive materials concurrently. This approach allows for the simultaneous and large-scale manipulation of single cells, while using lower light energy levels, enabling experimentation over a broader processing area. Implementing this technology necessitates the integration of optical components and compatible chip assemblies [54].

The OEP assembly comprises an optical system and optically sensitive materials. The light source is directed to illuminate the target area through a lens. As the incident light interacts with a hydrogenated silicon photoconductive electrode, it generates a localized electric field above the illuminated region, resulting in an electric field force. By manipulating the position of the incident light, particles at that specific location, such as cells, can be propelled. This OEP technology, when integrated with microfluidics external to the microchamber, facilitates the precise sorting and collection of live cells without the need for fluorescent dye labeling. The cells are directed into nanoliter-scale culture chambers called Nano-Pens, enabling their cultivation and analysis in a controlled environment.

The Beacon® instrument was developed by Berkeley Lights [54, 55]. The microfluidic device from Berkeley Lights features up to 14,000 "NanoPen" chambers and microfluidic channels that interconnect all "NanoPen" chambers , facilitating the loading/extraction of cells, beads, and the delivery of reagents. The NanoPens have a volume capacity ranging from 250 pL to 2 nL, depending on the intended use. These minute volumes facilitate the rapid examination of individual cell phenotypes. As shown in Figure 6.23a, Berkeley Lights' OEP technology is used to

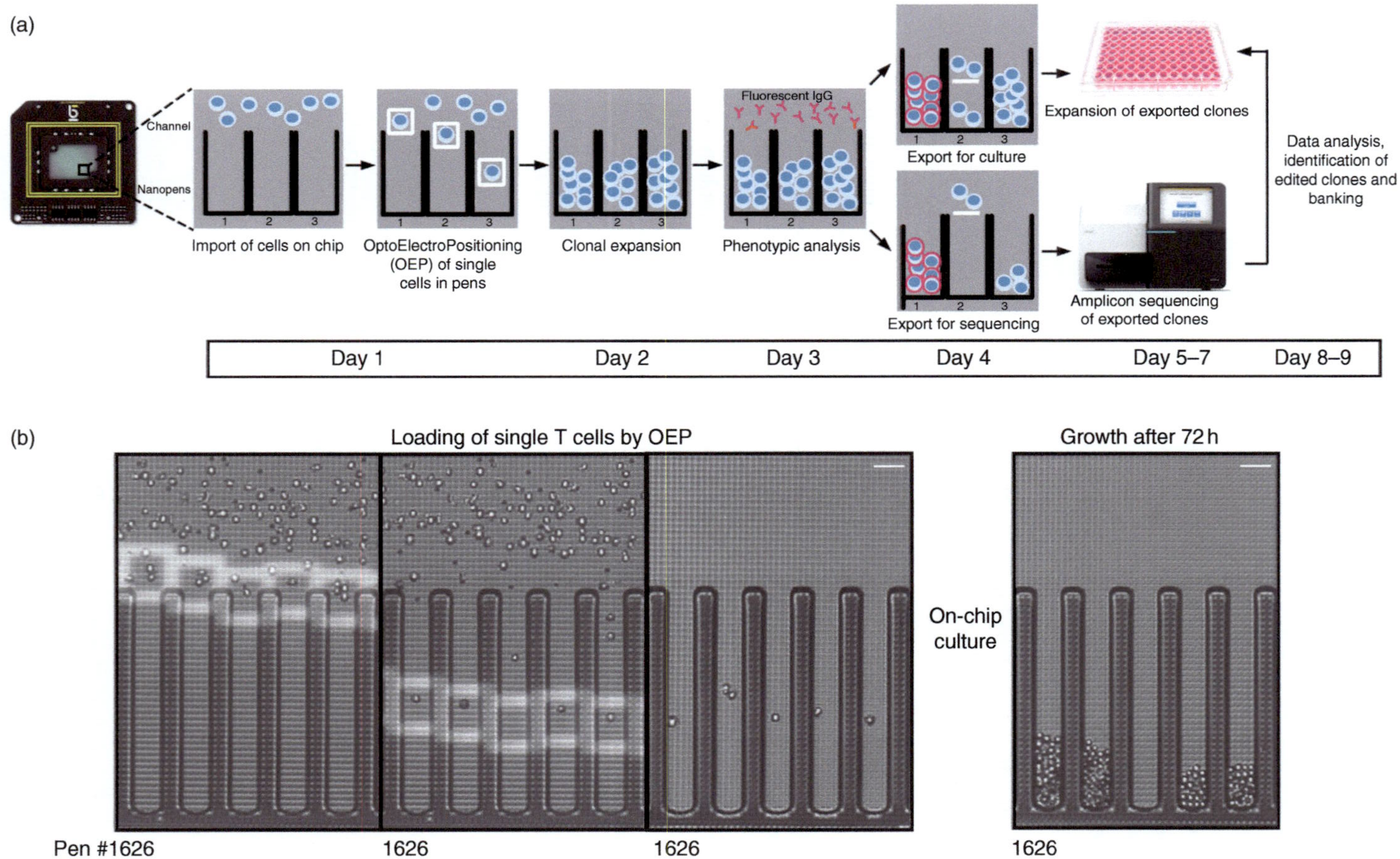

Figure 6.23 (a) Berkeley Lights' Beacon system for cell screening and selection; (b) electroporated cells in the channel are auto-identified, captured in light cages. *Source*: [52] / Springer Nature / CC BY 4.0.

activate single T cells, which are then grown into clones after 72 hours of cultivation. Figure 6.23b details the standard procedure for cell analysis and selection with the Beacon platform [51, 52]. The process begins with cells being loaded onto the OptoSelect chip, which combines OET with the microfluidic technology. Cells are directed into the NanoPens. Once the clonal growth is achieved, specific reagents are applied to the chip to assess various phenotypic characteristics, including protein secretion, cytokine presence, and cell surface markers. Following characterization, selected cell colonies are removed for the off-chip analysis, which may involve genomic studies and additional amplification processes in microplates. This comprehensive approach lets researchers accurately link individual cell phenotypes to their genetic information.

The Beacon system's versatility allows it to be used for a range of applications, from direct antibody candidate discovery using B cells to creating effective T cell therapies [47]. Compared to conventional methods such as hybridoma technology, which requires 2–3 months to isolate antibodies from B cells, the Beacon platform dramatically shortens this timeline to a few days, significantly speeding up the creation and development of cell-based products [49]. The system is more resource-efficient, using fewer cells and reagents than traditional methods. Consequently, users can quickly and cost-effectively identify optimal clones, making the Beacon® an essential asset for fast-tracking antibody studies and product development. Recently, the Beacon® system's adaptability and rapid processing capabilities have been employed to identify potential treatments and vaccines for COVID-19 [56–58]. Growing numbers of research entities and biopharmaceutical firms globally are adopting the platform. To better cater to the research and academic sectors, which may not require high-throughput automation, Berkeley Lights has also introduced a smaller, more affordable system called Lightning. In summary, leveraging OET as its foundational technology, Berkeley Lights is well-positioned to continue impacting the development of cell therapies, biopharmaceuticals, and other cell-based innovations.

References

1 Zhao, J., Gan, C., Zhang, J. et al. (2022). Deep learning assisted automated separation platform of single cells and microparticles using optoelectronic tweezers. *WRC Symposium on Advanced Robotics and Automation (WRC SARA)* 2022: 275–279.

2 Ke, L.Y., Kuo, Z.K., Chen, Y.S. et al. (2017). Cancer immunotherapy mu-environment LabChip: taking advantage of optoelectronic tweezers. *Lab on a Chip* 18 (1): 106–114.

3 Huang, K.W., Wu, Y.C., Lee, J.A., and Chiou, P.Y. (2013). Microfluidic integrated optoelectronic tweezers for single-cell preparation and analysis. *Lab on a Chip* 13 (18): 3721–3727.

4 Chiu, T.K., Chou, W.P., Huang, S.B. et al. (2016). Application of optically-induced-dielectrophoresis in microfluidic system for purification of circulating tumour cells for gene expression analysis- cancer cell line model. *Scientific Reports* 6: 32851.

5 Liao, C.J., Hsieh, C.H., Chiu, T.K. et al. (2018). An optically induced dielectrophoresis (ODEP)-based microfluidic system for the isolation of high-purity CD45(neg)/EpCAM(neg) cells from the blood samples of cancer patients-demonstration and initial exploration of the clinical significance of these cells. *Micromachines (Basel)* 9 (11): 563.
6 Huang, S.B., Wu, M.H., Lin, Y.H. et al. (2013). High-purity and label-free isolation of circulating tumor cells (CTCs) in a microfluidic platform by using optically-induced-dielectrophoretic (ODEP) force. *Lab on a Chip* 13 (7, 7): 1371–1383.
7 Chou, W.-P., Wang, H.-M., Chang, J.-H. et al. (2017). The utilization of optically-induced-dielectrophoresis (ODEP)-based virtual cell filters in a microfluidic system for continuous isolation and purification of circulating tumour cells (CTCs) based on their size characteristics. *Sensors and Actuators B: Chemical* 241: 245–254.
8 Liang, W., Zhao, Y., Liu, L. et al. (2014). Rapid and label-free separation of Burkitt's lymphoma cells from red blood cells by optically-induced electrokinetics. *PLoS One* 9 (6): e90827.
9 Chiou, P.Y., Ohta, A.T., and Wu, M.C. (2005). Massively parallel manipulation of single cells and microparticles using optical images. *Nature* 436 (7049): 370–372.
10 Ohta, A.T., Chiou, P.Y., Phan, H.L. et al. (2007). Optically controlled cell discrimination and trapping using optoelectronic tweezers. *IEEE Journal of Selected Topics in Quantum Electronics* 13 (2): 162–167.
11 Hwang, H., Lee, D.H., Choi, W., and Park, J.K. (2009). Enhanced discrimination of normal oocytes using optically induced pulling-up dielectrophoretic force. *Biomicrofluidics* 3 (1): 14103.
12 Ohta, A.T., Garcia, M., Valley, J.K. et al. (2010). Motile and non-motile sperm diagnostic manipulation using optoelectronic tweezers. *Lab on a Chip* 10 (23): 3213–3217.
13 Chu, P.-Y., Liao, C.-J., Hsieh, C.-H. et al. (2019). Utilization of optically induced dielectrophoresis in a microfluidic system for sorting and isolation of cells with varied degree of viability: demonstration of the sorting and isolation of drug-treated cancer cells with various degrees of anti-cancer drug resistance gene expression. *Sensors and Actuators B: Chemical* 283: 621–631.
14 Liang, W., Liu, J., Yang, X. et al. (2020). Microfluidic-based cancer cell separation using active and passive mechanisms. *Microfluidics and Nanofluidics* 24 (4).
15 Miccio, L., Marchesano, V., Mugnano, M. et al. (2016). Light induced DEP for immobilizing and orienting bacteria. *Optics and Lasers in Engineering* 76: 34–39.
16 Wang, H.Y., Chen, C.Y., Chu, P.Y. et al. (2020). Application of an optically induced dielectrophoresis (ODEP)-based microfluidic system for the detection and isolation of bacteria with heterogeneity of antibiotic susceptibility (vol 307, 127540, 2020). *Sensors and Actuators B: Chemical* 323.
17 Chen, Y.S., Lai, C.P.K., Chen, C.C., and Lee, G.B. (2021). Isolation and recovery of extracellular vesicles using optically-induced dielectrophoresis on an integrated microfluidic platform. *Lab on a Chip* 21 (8): 1475–1483.

18 Yang, S.M., Tseng, S.Y., Chen, H.P. et al. (2013). Cell patterning via diffraction-induced optoelectronic dielectrophoresis force on an organic photoconductive chip. *Lab on a Chip* 13 (19): 3893–3902.
19 Gan, C., Xiong, H., Zhao, J.et al (2023). Parallel cell array patterning and target cell lysis on an optoelectronic micro-well device. 2023 IEEE/RSJ International Conference on Intelligent Robots and Systems (IROS) (1–5 October), IEEE, pp. 2635–2640.
20 Yu, T.-M., Yang, S.-M., Huang, H.-P. et al. (2010). Dynamic manipulation and patterning of micro particles and cells by using TiOPc-based optoelectronic dielectrophoresis. *Optics Letters* 35 (12): 1959–1961.
21 Hu, W., Ishii, K., Ohta, A.T.2016). Cell patterning in a hydrogel using optically induced dielectrophoresis. 2016 International Conference on Optical MEMS and Nanophotonics (OMN), Singapore (31 July–4 August), IEEE, 2016.
22 Lin, Y.H., Yang, Y.W., Chen, Y.D. et al. (2012). The application of an optically switched dielectrophoretic (ODEP) force for the manipulation and assembly of cell-encapsulating alginate microbeads in a microfluidic perfusion cell culture system for bottom-up tissue engineering. *Lab on a Chip* 12 (6): 1164–1173.
23 Gi, H.J., Han, D., and Park, J.K. (2017). Optoelectrofluidic printing system for fabricating hydrogel sheets with on-demand patterned cells and microparticles. *Biofabrication* 9 (1): 015011.
24 Yang, W., Yu, H., Li, G. et al. (2017). High-throughput fabrication and modular assembly of 3D heterogeneous microscale tissues. *Small* 13 (5): 1602769.
25 Cui, J., Wang, H.P., Shi, Q., and Sun, T. (2021). Pulsed microfluid force-based on-chip modular fabrication for liver lobule-like 3D cellular models. *Cyborg and Bionic Systems* 2021: 9871396.
26 Yang, W.G., Cai, S.X., Chen, Y.B. et al. (2020). Modular and customized fabrication of 3D functional microgels for bottom-up tissue engineering and drug screening. *Advanced Materials Technologies* 5 (5).
27 Liu, N., Liang, W., Liu, L. et al. (2014). Extracellular-controlled breast cancer cell formation and growth using non-UV patterned hydrogels via optically-induced electrokinetics. *Lab on a Chip* 14 (7): 1367–1376.
28 Wang, S., Liang, W.F., Dong, Z.L. et al. (2011). Fabrication of micrometer- and nanometer-scale polymer structures by visible light induced dielectrophoresis (DEP) force. *Micromachines* 2 (4): 431–442.
29 Liu, N., Li, P., Liu, L. et al. (2015). 3-D non-UV digital printing of hydrogel microstructures by optically controlled digital electropolymerization. *Journal of Microelectromechanical Systems* 24 (6): 2128–2135.
30 Li, P., Yu, H.B., Liu, N. et al. (2018). Visible light induced electropolymerization of suspended hydrogel bioscaffolds in a microfluidic chip. *Biomaterials Science* 6 (6): 1371–1378.
31 Wang, H.Y.F., Liu, N., Mai, J.D. et al. (2013). Non-ultraviolet-based patterning of polymer structures by optically induced electrohydrodynamic instability. *Applied Physics Letters* 103 (21): 214101–214105.
32 Wang, F., Liu, L., Li, G. et al. (2018). Thermometry of photosensitive and optically induced electrokinetics chips. *Microsystems & Nanoengineering* 4: 26.

33 Gan, C., Zhang, J., Chen, B. et al. (2024). Optoelectronic tweezers micro-well system for highly efficient single-cell trapping, dynamic sorting, and retrieval. *Small* 20 (23): e2307329.

34 Jeorrett, A.H., Neale, S.L., Massoubre, D. et al. (2014). Optoelectronic tweezers system for single cell manipulation and fluorescence imaging of live immune cells. *Optics Express* 22 (2): 1372–1380.

35 Han, D. and Park, J.K. (2016). Optoelectrofluidic enhanced immunoreaction based on optically-induced dynamic AC electroosmosis. *Lab on a Chip* 16 (7, 7): 1189–1196.

36 Han, D. and Park, J.K. (2016). Microarray-integrated optoelectrofluidic immunoassay system. *Biomicrofluidics* 10 (3): 034106.

37 Hyundoo Hwang, H.C., Choo, J., and Park, J.-K. (2010). Optoelectrofluidic sandwich immunoassays for detection of human tumor marker using surface-enhanced Raman scattering. *Analytical Chemistry* 82 (18): 7603–7610.

38 Han, D., Hwang, H., and Park, J.-K. (2013). Optoelectrofluidic behavior of metal–polymer hybrid colloidal particles. *Applied Physics Letters* 102 (5): 054105.

39 Liang, W., Yang, X., Wang, J. et al. (2020). Label-free characterization of different kinds of cells using optoelectrokinetic-based microfluidics. *Optics Letters* 45 (8): 2454–2457.

40 Liang, W., Zhang, K., Yang, X. et al. (2015). Distinctive translational and self-rotational motion of lymphoma cells in an optically induced non-rotational alternating current electric field. *Biomicrofluidics* 9 (1): 014121.

41 Yang, X., Niu, X., Liu, Z. et al. (2017). Accurate extraction of the self-rotational speed for cells in an electrokinetics force field by an image matching algorithm. *Micromachines (Basel)* 8 (9).

42 Walid Rezanoor, M. and Dutta, P. (2016). Combined AC electroosmosis and dielectrophoresis for controlled rotation of microparticles. *Biomicrofluidics* 10 (2): 024101.

43 Zhu, X. (2015). Manipulation of self-assembled microparticle chains by electroosmotic flow assisted electrorotation in an optoelectronic device. *Micromachines* 6 (9): 1387–1405.

44 Zhao, Y., Lai, H.S., Zhang, G. et al. (2014). Rapid determination of cell mass and density using digitally controlled electric field in a microfluidic chip. *Lab on a Chip* 14 (22): 4426–4434.

45 Zhao, Y., Lai, H.S., Zhang, G. et al. (2015). Measurement of single leukemia cell's density and mass using optically induced electric field in a microfluidics chip. *Biomicrofluidics* 9 (2): 022406.

46 Kremer, C., Witte, C., Neale, S.L. et al. (2014). Shape-dependent optoelectronic cell lysis. *Angewandte Chemie (International Ed. in English)* 53 (3): 842–846.

47 Bruker. "Bruker Cellular Analysis." https://www.berkeleylights.com/ (accessed 12 January 2025).

48 Winters, A., McFadden, K., Bergen, J. et al. (2019). Rapid single B cell antibody discovery using nanopens and structured light. *MAbs* 11 (6): 1025–1035.

49 Jorgolli, M., Nevill, T., Winters, A. et al. (2019). Nanoscale integration of single cell biologics discovery processes using optofluidic manipulation and monitoring. *Biotechnology and Bioengineering* 116 (9): 2393–2411.

50 Le, K., Tan, C., Gupta, S. et al. (2018). A novel mammalian cell line development platform utilizing nanofluidics and optoelectro positioning technology. *Biotechnology Progress* 34 (6): 1438–1446.

51 Le, K., Tan, C., Le, H. et al. (2020). Assuring clonality on the beacon digital cell line development platform. *Biotechnology Journal* 15 (1): e1900247.

52 Mocciaro, A., Roth, T.L., Bennett, H.M. et al. (2018). Light-activated cell identification and sorting (LACIS) for selection of edited clones on a nanofluidic device. *Communications Biology* 1: 41.

53 Bronevetsky, Y. (2020). Directly test individual T cell function with fewer cells on the Berkeley lights lightning™ platform. *Cytotherapy* 22: S119.

54 Wu, M. C. 2019). Optoelectronic tweezers – an optofluidic platform for digital cell biology. 2019 International Conference on Optical MEMS and Nanophotonics (OMN) (28 July 2019), IEEE.

55 Rienzo, M., Lin, K.C., Mobilia, K.C. et al. (2021). High-throughput optofluidic screening for improved microbial cell factories via real-time micron-scale productivity monitoring. *Lab on a Chip* 21 (15): 2901–2912.

56 Bruker. (2023). "Published in Cell Reports: How Bruker Cellular Analysis' Beacon® Optofluidic System is Helping Researchers Stay a Step Ahead of COVID-19 Variants". https://www.brukercellularanalysis.com/blog (accessed 15 June 2024).

57 Zost, S.J., Gilchuk, P., Chen, R.E. et al. (2020). Rapid isolation and profiling of a diverse panel of human monoclonal antibodies targeting the SARS-CoV-2 spike protein. *Nature Medicine* 26 (9): 1422–1427.

58 Cho, H., Gonzales-Wartz, K.K., Huang, D. et al. (2021). Bispecific antibodies targeting distinct regions of the spike protein potently neutralize SARS-CoV-2 variants of concern. *Science Translational Medicine* 13 (616): eabj5413.

7

Applications in Physics and Materials Science

OETs can be used to manipulate various micro and nanoscale objects. At the same time, OETs can control a wide variety of objects, including living organisms (various types of cells [1–7], bacteria [8, 9], and microalgae [10, 11]) and nonliving organisms (various types of microspheres [12], nanoparticles [13, 14], hydrogel microstructures [15], and metal microrobots [16]). This chapter introduces in detail the application of OETs in manipulating micro and nano objects.

7.1 Manipulation of Microparticles

The most common objects that OETs operate on are various types of microspheres. Microspheres are easy to manufacture, have good uniformity, and have many applications in the biomedical field.

As shown in Figure 7.1a, Liang et al. [12] presented a novel method for the automated and parallel manipulation, separation, concentration, and assembly of micro- and nano-entities, employing OETs. The research highlights the use of the ODEP technology to effectively distinguish and concentrate polystyrene beads of various diameters by adjusting the directionality and magnitude of the ODEP forces. It begins with a theoretical analysis of the crossover frequency spectrum characteristics of polystyrene beads placed in a specific fluidic medium. Experimentation validated these theoretical predictions, demonstrating that micro- and nano-sized particles could be effectively manipulated using an OET chip.

In the experiments, researchers successfully achieved the rapid separation of beads with diameters of 10 μm and 1 μm by employing a combination of negative and positive ODEP forces in a medium with a conductivity of 1×10^{-3} S/m (Figure 7.1b). Additionally, using a higher liquid conductivity of 9×10^{-3} S/m, the team managed to manipulate and separate polystyrene beads with diameters of 500 nm, 1 μm, and 10 μm by projecting a shrinking concentric rectangular virtual electrode onto the optically induced electrokinetics (OEK) chip.

This technique is highly applicable in fields requiring the sorting and assembly of micro- and nanoscale particles, such as in biomedical applications and the fabrication of microscale devices.

Optoelectronic Tweezers for Manipulation at the Micro and Nano Scale, First Edition.
Lin Feng, Menglu Tan, Ao Wang, and Jiaying Zhang.
© 2026 WILEY-VCH GmbH. All rights reserved, including rights for text and data mining and training of artificial intelligence technologies or similar technologies. Published 2026 by WILEY-VCH GmbH.

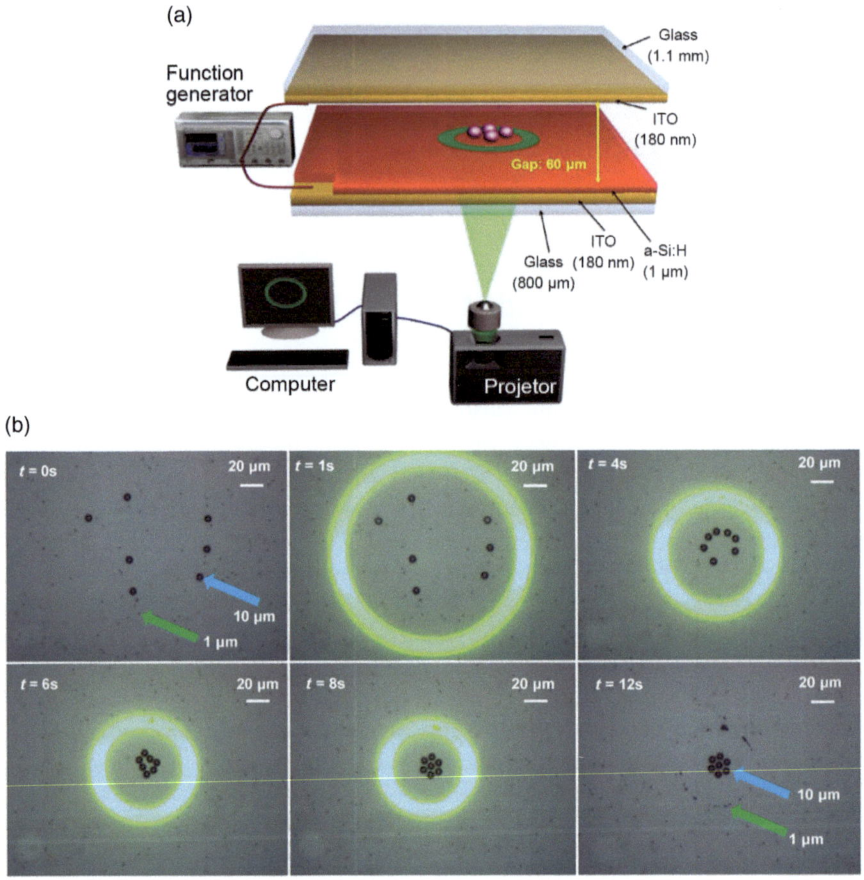

Figure 7.1 (a) Illustration of the trapping of particles by an image projected onto an OET chip [12]. (b) Dynamic separation of polystyrene beads with diameters of 10 and 1 µm, respectively. *Source*: [12] / with permission of Elsevier.

Liang et al. explored the assembly of microparticles using OETs, focusing on the interaction between particles with positive and negative dielectric properties [11] (Figure 7.2a, b). Contrasting with traditional OET systems that necessitate complex photoelectric material assemblies, this method utilizes a wireless, iPad-controlled platform for real-time manipulation, offering a more streamlined and contactless approach.

The research involved simulating interactions between micro-objects with positive and negative dielectric properties using the arbitrary Lagrangian–Eulerian (ALE) method, providing key insights into the factors affecting particle behavior in an electric field. This novel OET platform was then applied to demonstrate the interaction of swimming microorganisms, which possess positive dielectric properties, with microparticles exhibiting negative dielectric properties. A unique feature of this method allows a single particle to interact simultaneously with up to five others, significantly enhancing the high-throughput transportation capabilities of the OET system and enabling the assembly of distinctive microstructures.

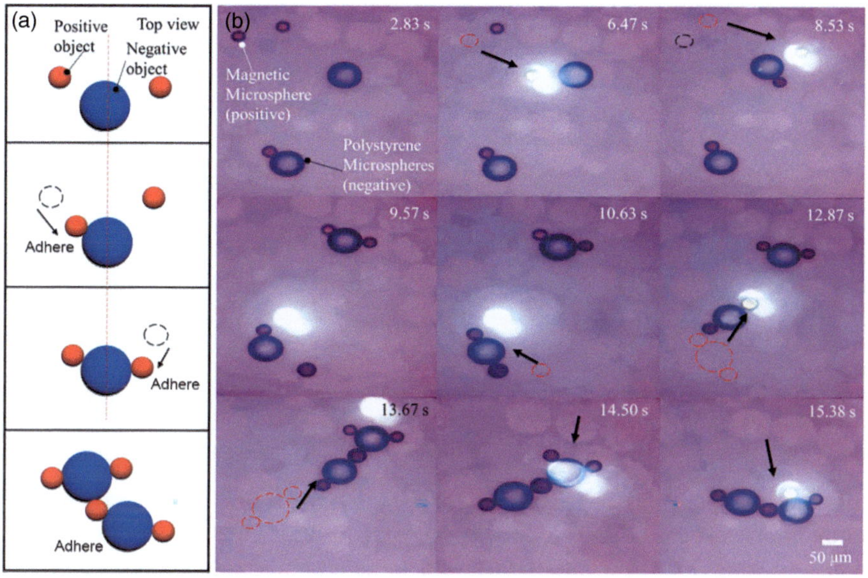

Figure 7.2 Interaction between a controllable positive-dielectric micro-object and multiple negative-dielectric microparticles using OET [11]. (a) Schematic of the interaction between polystyrene microbeads and magnetic particles; (b) polystyrene microbeads interact with magnetic particles. *Source*: [11] / The Royal Society of Chemistry.

One of the pivotal advantages of this low-power-intensity system is its ability to reduce experimental contamination and prevent adverse effects on the microorganisms involved, such as polystyrene microbeads used in transport experiments. The results indicate that the micro-objects, regardless of their complex shapes – spherical, rod-shaped, or spiral – were effectively manipulated without damage, showcasing the system's versatility.

Janus microparticles exhibit a unique dual-faced structure, with distinct physical or chemical properties on each side. This structure is achieved by coating one hemisphere of polystyrene spheres with a thin metallic layer, such as Cr, Ni, or Au. This coating enables the particles to exhibit self-propelling behavior when subjected to an AC electric field. The propulsion mechanisms involved include induced-charge electrophoresis (ICEP) and self-dielectrophoresis (sDEP), setting Janus particles apart from traditional microparticles that rely on external forces for movement (Figure 7.3a, b).

Janus particles can be precisely manipulated using OET, allowing for the formation of intricate trajectories and assemblies that are challenging to achieve with conventional particles [17]. Researchers have demonstrated the ability to control the propulsion mode of these particles by adjusting the frequency of the electric field and the shape of optical patterns. This control enables the particles to self-assemble into stable configurations or follows complex paths.

The real-time reconfigurability of the Janus particle movement and assembly using OET provides significant advantages over traditional manipulation methods. This capability is especially valuable in applications requiring precise control over

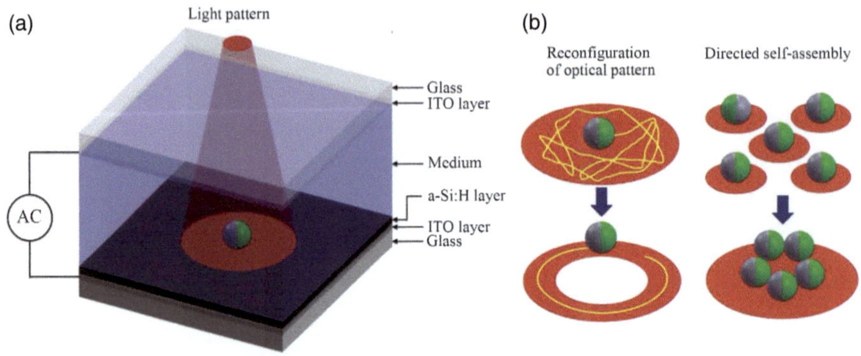

Figure 7.3 (a) Diagram of the microfluidic chamber used for the optoelectronic (OE) manipulation of active particles. (b) Adjustment of active particle trajectories and direct self-assembly of active particles [17]. *Source*: [17] / John Wiley & Sons.

particle behavior, such as targeted drug delivery, microrobotics, and environmental cleanup efforts. These studies highlight the versatility of OET in handling Janus particles and suggest its potential to transform the manipulation and utilization of microscale materials in various scientific and industrial domains.

Liang et al. [13] proposed an application of OET for the manipulation and assembly of carbon nanoparticles (CNPs) into functional electrical components within a specially designed microfluidics chip. By utilizing OETs, this study addresses the challenge of controlling CNPs through optically induced electroosmotic flow (OIEF), which is crucial for overcoming the randomizing effects of Brownian motion that typically impede nanoparticle manipulation.

The experiment demonstrated the practicality of using OET to assemble CNPs into well-defined three-dimensional microstructures. The process utilizes a combination of incident light and an externally applied AC bias potential to create optically defined virtual electrodes within the microfluidic environment. This innovative setup allowed for the CNPs to be precisely manipulated and assembled into specific configurations within approximately 45 seconds – a speed that underscores the efficiency of the technique.

The proven capability of the OET-enabled OIEF technique to rapidly assemble and tailor CNP-based microstructures suggests its potential for broader applications in electronics. This method could facilitate the integration of various nanoparticle elements into complex electronic devices, paving the way for advanced microelectronic systems and innovative applications in fields such as sensors, actuators, and other microelectromechanical systems (MEMS).

7.2 Manipulation of Nanoparticles

7.2.1 Nanoparticles

Liu et al. [14] proposed an application of OET for the single-step fabrication and surface modification of microelectrodes from gold nanoparticles (AuNPs) using an

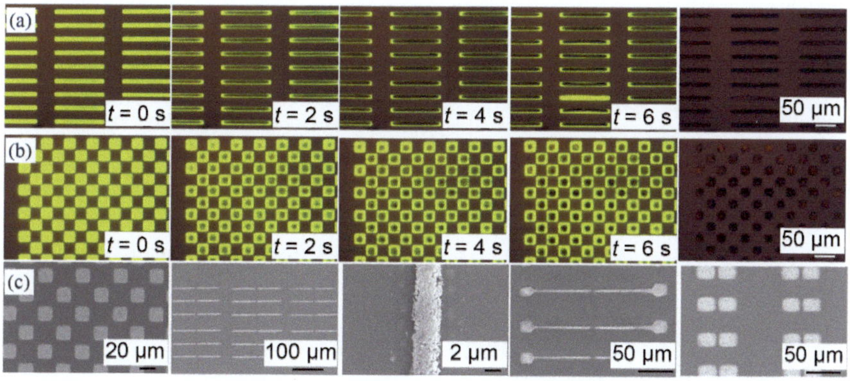

Figure 7.4 (a, b) Patterning process and (c) SEM images of gold electrodes showing optical and deposition results. *Source*: [14] / IOP Publishing.

optically induced electrodeposition (OED) method (Figure 7.4a–c). This innovative technique has garnered significant attention due to its potential applications in chemistry and biomedical engineering, including surface-enhanced Raman spectroscopy (SERS) and catalytic sites for electrochemistry. The OED process leverages the principles of OET to rapidly deposit and pattern gold electrodes and simultaneously modify their surfaces with nanoparticles within just 6 seconds.

The researchers successfully fabricated and modified arrays of gold electrodes with customized shapes within a 500 μm diameter area. The conductivity of these electrodes ranged from 6×10^5 to 8×10^5 S/m, and the electrodes could be transferred onto flexible substrates like polydimethylsiloxane (PDMS). This capability underscores the method's versatility in utilizing substrates beyond a-Si:H. The study demonstrated that the frequency and amplitude of the applied AC signal, along with deposition time, are critical parameters influencing the electrode morphology. The weak adhesion of thicker electrodes to the a-Si:H substrate facilitates their transfer to other substrates, expanding their potential applications.

7.2.2 Multiwalled Carbon Nanotubes

OET enables the precise trapping and manipulation of multiwalled carbon nanotubes (MWCNTs). Through the projection of optical patterns onto a photoconductive surface, localized electric fields are created, facilitating the manipulation of individual nanotubes with low optical power requirements. This approach allows for high-speed translation (200 μm/s) and two-dimensional trapping of MWCNTs, providing notable benefits in terms of parallel processing capabilities [18] (Figure 7.5a–e). The precise manipulation of MWCNTs through OET holds promise for various applications in the nanomaterial assembly, sensor development, and nanocomposite fabrication. By harnessing light patterns to generate dielectrophoretic forces, the deposition of the polyaniline PANI/MWCNT composite is controlled with accuracy, resulting in the formation of uniform and continuous electrodes. This method shows a great potential in fields like sensor development,

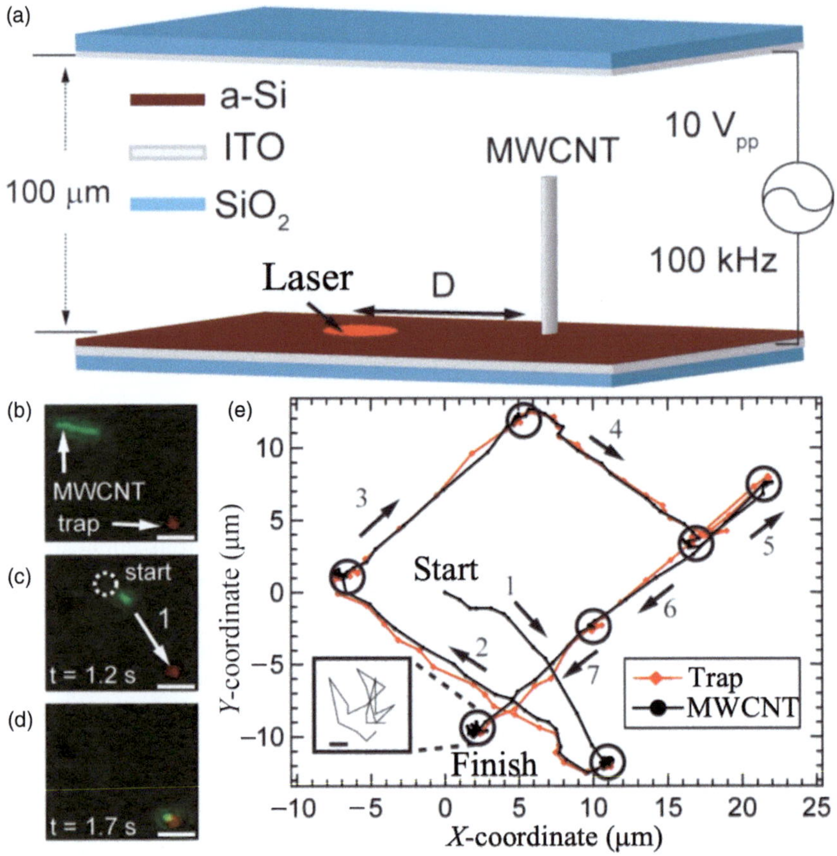

Figure 7.5 (a) OET chamber schematic. (b) Dark field image of MWCNT moving randomly. (c) Alignment and initial movement. (d) Translation of the MWCNT into the OET trap. (e) MWCNT and laser coordinates from video frames [18]. *Source*: [18] / Springer Nature / CC BY-NC-ND 3.0.

flexible electronics, and nanocomposite materials, offering an effective alternative to conventional lithographic techniques.

Liu et al. [19] introduced a method utilizing OET for the micro-patterning of conductive polymer/carbon nanotube composite electrodes, specifically a PANI/ MWCNT composite, using an OEK chip. This flexible and dynamically programmable technique is significant due to its low cost and potential applications across various domains. The method allows for the rapid and flexible patterning of polymeric electrodes with varying geometries by applying a square waveform signal with amplitudes ranging from 16 to 20 V and frequencies between 20 and 30 kHz.

The OEK-based method developed here eliminates the need for molds, as required by traditional micro-contact printing, hence simplifying the preparation process. While inkjet printing also avoids molds, the OEK-based approach offers superior pattern resolution through high numerical aperture projection optics, with theoretical limits determined only by diffraction. A significant advantage of this OEK-based technique over existing methods lies in its ease of reconfiguration and patterning flexibility. The method's ability to use customized light patterns and adjust exposure times allows for precise control over electrode shapes and dimensions. Future work should focus on enhancing the structural stability through post-processing and expanding the diversity of polymer materials amenable to this OEK-based patterning. This research demonstrates a significant advancement in the field of electrode fabrication, highlighting the versatility and efficiency of using OET in micro-patterning applications.

7.2.3 Nanowires

Jamshidi et al. [20] introduced a novel method utilizing OET for the precise manipulation and assembly of individual nanowires (Figure 7.6). This technique addresses the challenge of handling semiconducting and metallic nanowires with diameters below 20 nm, facilitating their use in integrated, heterogeneous optoelectronic devices at the nanoscale. Employing significantly lower optical power density – 100,000 times less than that of OT – OET enables the transportation of individual nanowires at speeds four times greater than the fastest speeds achieved by traditional OT.

In this study, a real-time array of silver nanowires is created using photopatterned virtual electrodes, illustrating the potential for massively parallel assemblies. OET also demonstrates capability in separating semiconducting from metallic nanowires, suggesting broad applications in the separation and heterogeneous integration of one-dimensional nanoscale materials.

Furthermore, the use of a DMD allows for large-scale assembly and dynamic trapping of nanowires. For instance, a 5 × 5 array of individually addressable silver nanowires was formed, and experiments demonstrated the adjustment of the nanowire density by varying the size of a trapping pattern in real-time. This process leverages dipole–dipole interactions to arrange nanowires within the active trapping pattern.

While this approach is promising for large-scale nanowire integration, current challenges include manipulation limitations to 2D surfaces and nonspecific adhesion of wires to sample surfaces. The surface functionalization of nanowires has been identified as a potential strategy to mitigate adhesion issues, with future prospects including the development of 3D heterostructures from metallic and semiconducting wires through lateral trapping and immobilization.

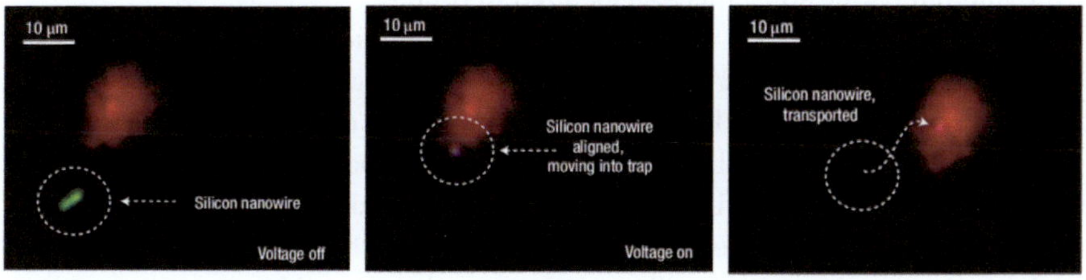

Figure 7.6 OET manipulation of single silicon nanowire [20]. *Source:* [20] / MDPI / CC BY 4.0.

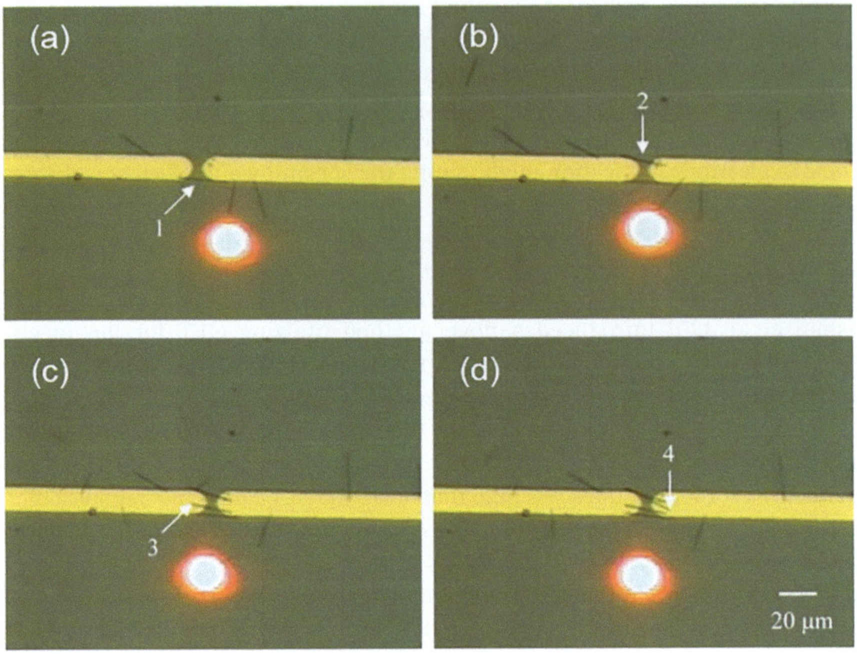

Figure 7.7 (a–d) The process connecting a single nanowire to multiple pairs of adjacent electrodes [21]. *Source*: [21] / Optical Society of America.

Overall, the real-time capabilities of OET offer parallel, high-speed manipulation for a wide range of one-dimensional materials, enabling the dynamic positioning and rapid separation of nanowires. The technique holds potential for extension into microfluidic chambers and could be applied to the separation of materials like carbon nanotubes or the construction of thermoelectric arrays.

As shown in Figure 7.7a–d, OET can be used to dynamically position nanowires near pre-patterned metal electrodes, followed by the application of an AC signal to the electrodes, generating a DEP force that precisely aligns and connects the nanowires [21].

This approach addresses the challenges in controlling the number and positioning of nanowires in sensor assembly. The ODEP platform uses optical images as virtual electrodes to activate a non-contact DEP force, which enables the manipulation of micron- and nanoscale substances suspended in fluid.

In this study, nanowires were first maneuvered adjacent to pre-deposited metal electrodes using virtual electrodes created by optical images. Subsequently, DEP forces were employed by activating alternating current signals to entice and align the nanowires into the gap between two electrodes. This technique allows a single nanowire to be assembled within 24 seconds, and it enables precise control over the number of nanowires in a single sensor. Furthermore, the assembly of an individual nanowire on each adjacent electrode is achievable.

Characterization of the assembled nanowires' electrical properties was performed through the I–V curve measurement. The study also investigated the contact resistance between nanowires and electrodes, as well as the adherence between nanowires and substrates. By patterning platinum at both ends of the nanowires post-interconnection, the setup effectively secured the nanowires and minimized movement during subsequent processes, reducing contact resistance and enhancing signal stability.

This OET-based method showcases significant advantages in operation flexibility and fast assembly. It permits the effective assembly of single nanowires to multiple pairs of electrodes and the integration of multiple nanowires to a single pair of electrodes. The process can successfully utilize a manipulating solution of pure DI water on an oxygen plasma-treated SiNx substrate, improving nanowire adherence issues. Effective manipulation was sustained at light intensities as low as 200 nW/μm^2, significantly lower than the requirement for OT.

The proposed method demonstrates a substantial increase in the assembly success rates compared to conventional methods, allowing for the precise control of nanowire quantity and presenting a versatile platform for nanowire sensor assembly. This flexible assembly method ensures the stable detection capabilities of sensors with appropriate insulation, even in wet environments, offering a robust solution for integrated nanoscale sensor development.

7.3 Manipulation of Droplet

Lee et al. introduced a cutting-edge optoelectrofluidic device that merges microfluidic channels for the creation and manipulation of droplets via OET [22] (Figure 7.8a, b). This device consists of a photoconductive layer, a microfluidic channel, and a ground electrode. The photoconductive layer is fabricated by depositing an ITO layer onto a glass substrate, followed by layers of amorphous silicon and silicon nitride. The microfluidic channel, constructed using PDMS, is selectively perforated to expose parts of the photoconductive surface, enabling droplet manipulation. The system is engineered to produce and manipulate water droplets within an oil medium, with droplet generation occurring at a T-shaped

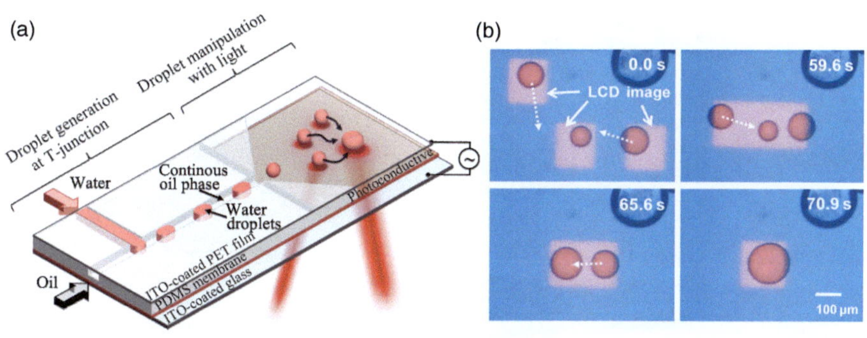

Figure 7.8 (a) Schematic illustration of an optoelectrofluidic device combined with a microfluidic channel. (b) Captured footage demonstrates interactive manipulation, including the transport and merging of multiple droplets via an LCD image [22]. *Source:* [22] / AIP Publishing.

microchannel where the flow rates of the water and oil phases are carefully controlled. These droplets are then transported to a manipulation region where they are directed using virtual electrodes formed by projecting the LCD images onto the photoconductive layer. The OET mechanism operates through positive DEP forces generated by the virtual electrodes, enabling versatile droplet manipulation activities such as transport, merging, and division.

Experimental findings demonstrated the device's capability to produce droplets of varying sizes, ranging from 0.57 to 4.9 nL, based on the flow rate ratios of the water and oil phases. The OET system enabled the precise and simultaneous manipulation of multiple droplets, with a maximum droplet movement velocity of 138.8 μm/s under a 40 V_{pp} voltage at 100 kHz. However, higher voltages impeded droplet movement due to increased frictional forces caused by electrowetting effects.

This device holds significant promise for high-throughput biological and chemical analyses, where the controlled manipulation of small-volume droplets is essential. The integration of microfluidic generation with OET-based manipulation in a unified platform offers a versatile tool for diverse lab-on-a-chip applications.

The ODEP droplet manipulation platform, as described by Park et al. and illustrated in Figure 7.9, is based on the concept of floating electrode optoelectronic tweezers (FEOETs). This technology leverages light patterns to create electric fields on a photoconductive surface, generating dielectrophoretic forces that enable the manipulation of aqueous droplets within an electrically non-conductive oil medium. The operational principle involves applying a DC bias across the photoconductive layer, with the light patterns serving as virtual electrodes. These virtual electrodes induce dipole interactions within the droplets, facilitating precise control over their movement. The platform offers the capability for the 2D manipulation of droplets,

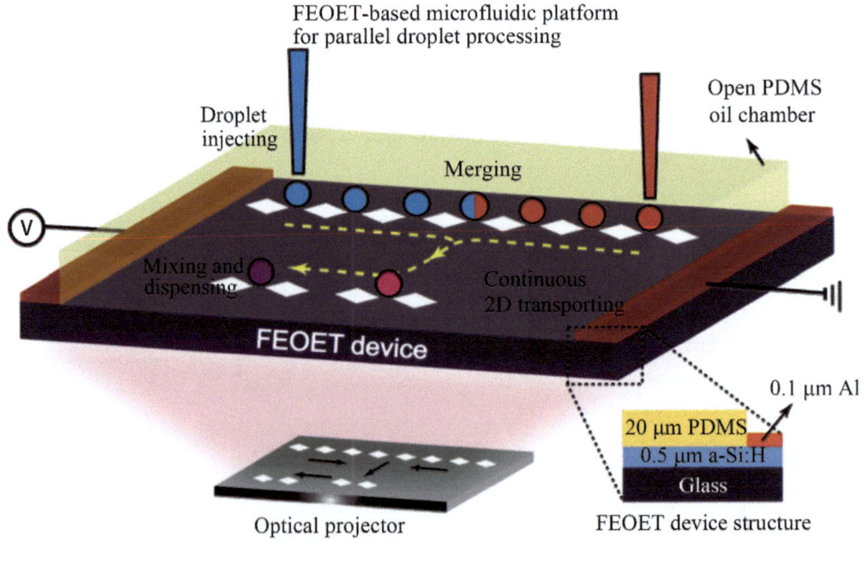

Figure 7.9 Schematic of a FEOET platform for parallel droplet manipulation using direct optical images [23]. *Source*: [23] / Royal Society of Chemistry.

encompassing tasks such as transport, merging, mixing, and parallel processing of multiple droplets [23].

The shape of the light pattern plays a crucial role in determining the direction and efficiency of droplet movement in the DEP droplet manipulation platform. While a circular laser beam can effectively move droplets parallel to the applied electric field, it may be less effective in perpendicular directions. To address this limitation, the study introduces paired-diamond (PD) shaped light patterns that enhance electric field gradients in both directions, enabling the effective two-dimensional manipulation of droplets.

This platform holds significant potential across various fields, particularly in biochemical and chemical analyses. It enables precise control over droplet movement, a critical requirement for applications such as enzyme kinetics, DNA amplification, and nanoparticle synthesis. The open chamber design of the FEOET platform allows for seamless integration with other microfluidic components, including microwell arrays and closed-channel systems, expanding its utility in lab-on-a-chip technologies. The platform's ability to manipulate droplets with precision and flexibility positions it as a valuable tool in advancing microfluidic systems.

7.4 Manipulation of Micromachines

How to power microstructures at the micro- and nanoscale has always been a difficult problem. As a field control technology, OET is a good source of power.

Yang et al. developed an advanced method combining optofluidic maskless lithography with OET to fabricate and manipulate photocurable hydrogel microstructures (Figure 7.10a–l). In this innovative approach, hydrogel microstructures are dynamically synthesized within a microfluidic device through UV-induced polymerization and subsequently transferred to an OET chip for precise manipulation. This method allows for the assembly of diverse microstructures into complex and reconfigurable 3D architectures. The OET system leverages a DMD to modulate light patterns, creating virtual electrodes that generate dielectrophoretic forces to move and assemble the hydrogel components [15]. The potential applications of this technique are broad and impactful. In microscale tissue engineering, the assembled microstructures can function as building blocks for constructing functional tissue architectures. In drug delivery systems, the engineered hydrogel structures could be designed to release therapeutic agents in a controlled and targeted manner. This versatile approach holds a significant promise for advancing personalized medicine and regenerative therapies, offering new avenues for the development of innovative biomedical applications.

In a recent study, Li et al. developed an innovative approach using OET to manipulate micro- and nanoscale structures, specifically focusing on the creation and operation of nano-kirigami microrobots [16]. These microrobots, inspired by the art of kirigami, were created through a precise fabrication process involving electron beam lithography and ion beam etching. This method resulted in intricate structures with reconfigurable three-dimensional shapes. The researchers successfully

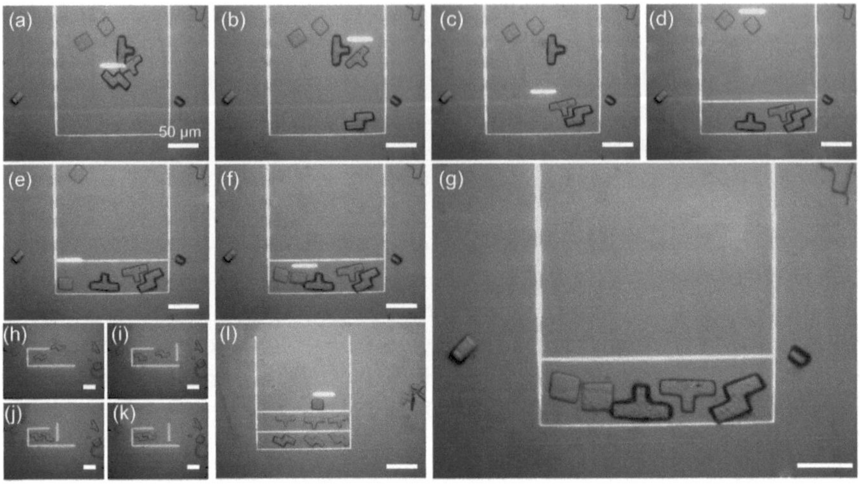

Figure 7.10 OET manipulates microstructures made from the in situ curing of photosensitive hydrogels [15]. (a–g) Sequential images of building a horizontal line with Tetris-like blocks; (h–k) two microstructures precisely aligned; (l) three block types layered. *Source*: [15] / John Wiley & Sons.

demonstrated the release and transfer of these nano-kirigami microrotors from their substrates, enabling their movement within microfluidic environments. The microrotors, distinguished by their elaborate kirigami designs, were manipulated using light patterns projected onto the OET system, allowing for customizable movement and positioning. The technology has a wide range of potential applications, including advanced microfluidic systems, nanophotonics, and micromechanical systems. These microrobots could serve as versatile and precise reconfigurable components in various fields, offering opportunities for innovative developments in microscale technologies.

References

1 Hwang, H., Lee, D.H., Choi, W., and Park, J.K. (2009). Enhanced discrimination of normal oocytes using optically induced pulling-up dielectrophoretic force. *Biomicrofluidics* 3 (1): 14103.

2 Hsu, H.Y., Ohta, A.T., Chiou, P.Y. et al. (2010). Phototransistor-based optoelectronic tweezers for dynamic cell manipulation in cell culture media. *Lab on a Chip* 10 (2): 165–172.

3 Song-Bin, H., Shing-Lun, L., Jian-Ting, L., and Min-Hsien, W. (2014). Label-free live and dead cell separation method using a high-efficiency optically-induced dielectrophoretic (ODEP) force-based microfluidic platform. *International Journal of Automation and Smart Technology* 4 (2): 83–91.

4 Zhao, Y., Lai, H.S., Zhang, G. et al. (2014). Rapid determination of cell mass and density using digitally controlled electric field in a microfluidic chip. *Lab on a Chip* 14 (22): 4426–4434.

5 Liang, W., Zhang, K., Yang, X. et al. (2015). Distinctive translational and self-rotational motion of lymphoma cells in an optically induced non-rotational alternating current electric field. *Biomicrofluidics* 9 (1): 014121.

6 Zhang, Y., Zhao, J., Yu, H. et al. (2020). Detection and isolation of free cancer cells from ascites and peritoneal lavages using optically induced electrokinetics (OEK). *Science Advances* 6 (32): eaba9628.

7 Gan, C., Zhang, J., Chen, B. et al. (2024). Optoelectronic tweezers micro-well system for highly efficient single-cell trapping, dynamic sorting, and retrieval. *Small* e2307329.

8 Wang, H.-Y., Chen, C.-Y., Chu, P.-Y. et al. (2020). Application of an optically induced dielectrophoresis (ODEP)-based microfluidic system for the detection and isolation of bacteria with heterogeneity of antibiotic susceptibility. *Sensors and Actuators B: Chemical* 307.

9 Chu, P.Y., Yang, C.M., Huang, K.L. et al. (2023). Development of an optically induced dielectrophoresis (ODEP) microfluidic system for high-performance isolation and purification of bacteria. *Biosensors-Basel* 13 (11).

10 Liang, S., Sun, J., Zhang, C. et al. (2022). Parallel manipulation and flexible assembly of micro-spiral via optoelectronic tweezers. *Frontiers in Bioengineering and Biotechnology* 10: 868821.

11 Liang, S., Gan, C., Dai, Y. et al. (2021). Interaction between positive and negative dielectric microparticles/microorganism in optoelectronic tweezers. *Lab on a Chip* 21 (22): 4379–4389.

12 Liang, W., Liu, N., Dong, Z. et al. (2013). Simultaneous separation and concentration of micro- and nano-particles by optically induced electrokinetics. *Sensors and Actuators A: Physical* 193: 103–111.

13 Liang, W., Liu, L., Wang, Y. et al. (2018). Rapid assembly of carbon nanoparticles into electrical elements by optically-induced electroosmotic flow. *IEEE Transactions on Nanotechnology* 17 (5): 1045–1052.

14 Liu, N., Li, M., Liu, L. et al. (2018). Single-step fabrication of electrodes with controlled nanostructured surface roughness using optically-induced electrodeposition. *Journal of Micromechanics and Microengineering* 28 (2).

15 Yang, W., Yu, H., Li, G. et al. (2017). High-throughput fabrication and modular assembly of 3D heterogeneous microscale tissues. *Small* 13 (5).

16 Hong, X., Xu, B., Li, G. et al. (2024). Optoelectronically navigated nano-kirigami microrotors. *Science Advances* 10 (17): eadn7582.

17 Das, S.S. and Yossifon, G. (2023). Optoelectronic trajectory reconfiguration and directed self-assembly of self-propelling electrically powered active particles. *Advanced Science* e2206183.

18 Pauzauskie, P.J., Jamshidi, A., Valley, J.K. et al. (2009). Parallel trapping of multiwalled carbon nanotubes with optoelectronic tweezers. *Applied Physics Letters* 95 (11): 113104.

19 Liu, N., Liang, W., Mai, J. D. et al. (2012). Rapid micro-patterning of a conductive PANI/MWNTs-polymer composite using an optically-induced electrokinetics chip. *In 2012 IEEE Nanotechnology Materials and Devices Conference (NMDC2012)* (10 June), IEEE, pp. 105–110.

20 Jamshidi, A., Pauzauskie, P.J., Schuck, P.J. et al. (2008). Dynamic manipulation and separation of individual semiconducting and metallic nanowires. *Nature Photonics* 2 (2): 86–89.

21 Lin, Y.H., Ho, K.S., Yang, C.T. et al. (2014). A highly flexible platform for nanowire sensor assembly using a combination of optically induced and conventional dielectrophoresis. *Optics Express* 22 (11): 13811–13824.

22 Lee, D.-H., Hwang, H., and Park, J.-K. (2009). Generation and manipulation of droplets in an optoelectrofluidic device integrated with microfluidic channels. *Applied Physics Letters* 95 (16).

23 Park, S.Y., Kalim, S., Callahan, C. et al. (2009). A light-induced dielectrophoretic droplet manipulation platform. *Lab on a Chip* 9 (22): 3228–3235.

8

Integration with Other Technologies

OETs are often combined with other technologies to accomplish a wider range of tasks. For example, they are combined with microfluidics for particle manipulation, with the photocuring technology for controlling hydrogel structures, with the electrowetting technology for droplet manipulation, and with the image recognition technology for more precise control. This chapter showcases these applications through specific examples.

8.1 Integration with Microfluidics

Chen et al. [1] presented a groundbreaking method for isolating and recovering extracellular vehicles (EVs) using a technique that integrates the OET technology with the microfluidic technology. This approach stands out for its ability to operate in a label-free and contact-free manner, which simplifies the traditional requirements for labeling and reduces the need for extensive manual intervention in the process (Figure 8.1).

The developed optically-induced EV (OIEV) platform uses innovative optical manipulation through virtual electrodes to facilitate a more efficient and effective management of EVs. The system operates under carefully optimized conditions, employing a 20 kHz alternating current frequency combined with a 20 V_{pp} driving voltage. A narrow 75-μm-width light pattern is utilized to generate an optically-induced dielectrophoretic (ODEP) force ranging from 23.5 to 97.7 fN in a 0.2 M sucrose solution. This precise setup allows for the rapid enrichment of EVs, resulting in a 272-fold concentration increase within a minute sample volume of 27 pL in just 32 seconds.

Beyond rapid concentration, the platform is designed to automatically isolate and recover EVs in approximately 25 minutes, boasting a remarkable release efficiency of 99.8% and a recovery rate of 52.2%. The label-free and contact-free characteristics of this approach help preserve the structural integrity of EVs, an essential factor for subsequent analytical and therapeutic applications. By reducing the necessity for manual handling, the platform enhances operational efficiency and scalability.

Additionally, the platform includes a circulating capture system that significantly improves the recovery process by ensuring the efficient enrichment

Optoelectronic Tweezers for Manipulation at the Micro and Nano Scale, First Edition.
Lin Feng, Menglu Tan, Ao Wang, and Jiaying Zhang.
© 2026 WILEY-VCH GmbH. All rights reserved, including rights for text and data mining and training of artificial intelligence technologies or similar technologies. Published 2026 by WILEY-VCH GmbH.

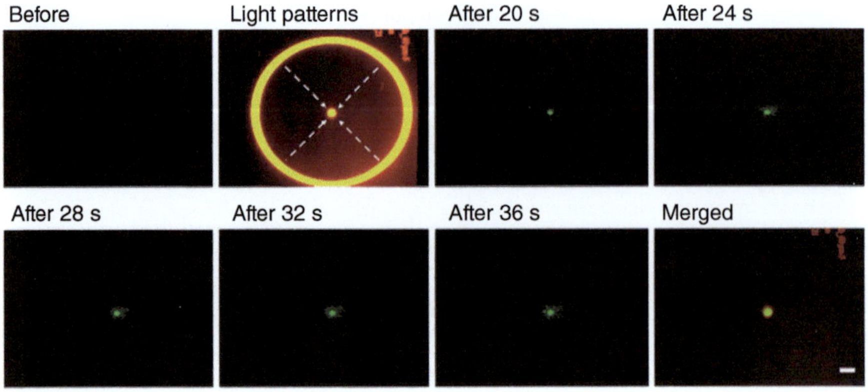

Figure 8.1 Enrichment of extracellular vesicles using the OET technology on a microfluidic platform. *Source*: [1] / The Royal Society of Chemistry.

and sorting of EVs. This capability is especially beneficial for applications in biomedicine and biotechnology, such as diagnostics and therapeutic delivery, where intact and unaltered EVs are crucial.

The potential applications of this integrated platform are vast. It shows significant promise in the realm of personalized medicine, particularly in the analysis of cancer-associated exosomal proteins and microRNAs, which are critical for early diagnosis and targeted therapies. The method's ability to handle EVs without altering their native state makes it an invaluable tool for exploring the role of EVs in various physiological and pathological processes.

In conclusion, this study demonstrates that the combination of ODEP and microfluidics offers a powerful and innovative solution for the isolation and recovery of EVs. By addressing the challenges associated with traditional EV handling techniques, this integrated approach paves the way for more robust and versatile applications in research and clinical settings, potentially revolutionizing how EVs are utilized across multiple disciplines in the biomedical field.

Pei et al. [2] developed a new micro–nano manipulation technique utilizing the OET technology and a digital microfluidic chip (Figure 8.2). They introduced a distributed circuit model specifically developed for multicolor light-actuated optoelectrowetting (OEW) devices. The model addresses key challenges in the OEW technology by considering the substantial variation in the absorption coefficient of photoconductors across the visible spectrum, which varies by a factor of 15. Additionally, the model considers the nonuniform distribution of photogenerated carriers within the photoconductive material, enabling a more accurate and efficient design of OEW devices.

Utilizing this model allowed the researchers to determine the optimal thickness of photoconductors for devices using multicolor light sources, such as those from commercial digital projectors. This optimization led to dramatic improvements in the device's performance. Specifically, the optical power requirement was reduced by 200 times compared to previous devices, making the system more energy-efficient. Additionally, the optimized device achieved a fivefold reduction in

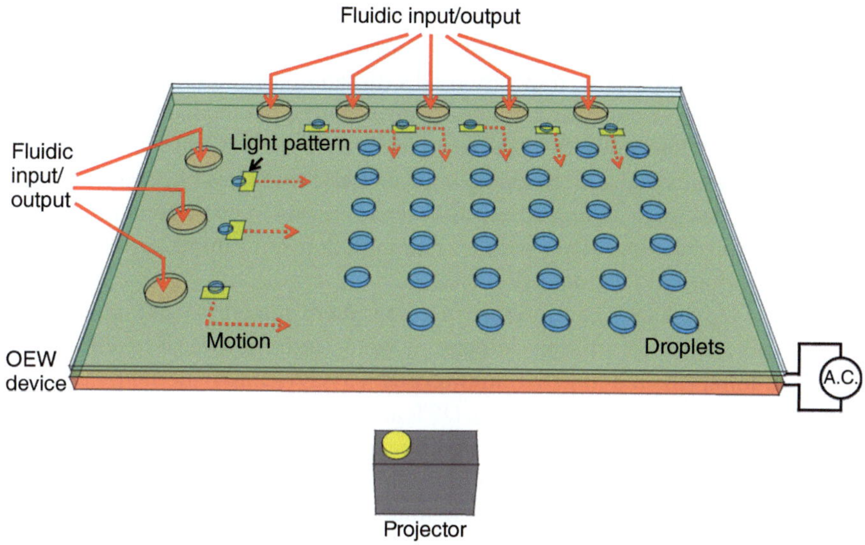

Figure 8.2 Schematic diagram of the structure and principle of the OET digital microfluidic chip. *Source*: [2] /Optica Publishing Group.

the voltage required for operation, enhancing the feasibility and safety of the OEW technology. Moreover, the droplet moving speed was increased by 20 times, achieving speeds greater than 1 cm/s, which is a significant enhancement for practical applications.

A key innovation highlighted in this study is the use of commercial digital projectors to create on-demand virtual electrodes for large-scale, parallel manipulation of droplets. This approach greatly reduces both the cost and complexity associated with light-actuated digital microfluidic systems, as it avoids the need for specialized and expensive light sources.

The platform demonstrated its capabilities by successfully manipulating a 96-droplet array simultaneously, showcasing its potential for high-throughput applications. Furthermore, the researchers demonstrated the platform's practical utility by performing the parallel on-chip detection of Herpes Simplex Virus Type 1 (HSV-1) within just 45 minutes through a real-time isothermal polymerase chain reaction (PCR) assay. This application underscores the platform's effectiveness for rapid and efficient biological assays, which could be vital for clinical diagnostics and research.

Overall, this light-actuated digital microfluidics platform offers significant advantages over traditional droplet manipulation methods. It provides enhanced versatility and flexibility due to its programmable nature, allowing for dynamic adjustments to meet specific experimental needs. The simplicity and low cost of the consumables further increase its accessibility and potential for widespread use. The use of readily available digital projectors highlights the practical applicability and potential for integration into existing laboratory setups, making this technology an appealing option for various scientific and industrial applications.

They utilized computer-aided design to project different shapes of light sources onto the chip, allowing for the flexible adjustment of electrode arrangements for various experiments. Additionally, they proposed a distributed circuit model that took into account color dependency and uneven distribution of photogenerated carriers. This model helped determine the optimal thickness of the light guide for a specific projector output spectrum, reducing optical power requirements significantly. This enabled the design of an optimized OEW device powered by a digital projector, achieving fast droplet movement, parallel manipulation of multiple droplets, and tasks such as isothermal PCR detection of HSV.

Cheng [3] developed a novel fluorescence microsphere detection technique based on a microwell array OET chip by integrating optoelectronic tweezer technology with the traditional enzyme-linked immunosorbent assay (ELISA) method. Initially, the principles of dielectrophoresis (DEP), OET, and ELISA detection were systematically analyzed and elucidated, with a two-dimensional visualization of the dielectrophoretic forces achieved through computer simulation software.

Subsequently, monodisperse polystyrene microspheres were utilized as carriers for the ELISA reaction. The corresponding specifications for the microwell array OET chip were manufactured using magnetron sputtering, photolithography, and microfluidic technology. Based on the principles of optoelectronic tweezer manipulation, an OET platform was designed and constructed to test the chip and select appropriate microspheres.

Prostate-specific antigen (PSA) was chosen as the target for detection experiments at various concentrations. The results were compared with those obtained using flow cytometry to serve as a control for the fluorescence microsphere detection technique developed in this study (Figure 8.3). The image recognition technology was employed for the automated analysis and processing of the experimental data. This design has effectively established a method for detecting proteins and other molecules in the microwell array OET chip. This approach offers advantages in terms of speed, flexibility, and cost over the ELISA detection techniques currently used in clinical settings.

Witte et al. [4] conducted a thorough investigation on the integration of OET technology with microfluidic systems to facilitate precise particle manipulation within microchannels (Figure 8.4a–c). The fabrication process commences with the creation of a microfluidic chip using the SU8 photoresist, which is patterned via

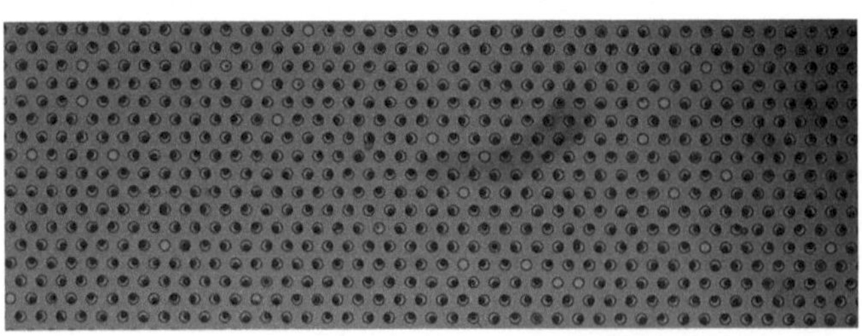

Figure 8.3 Two minutes after injection, the micro-well array achieves an adsorption rate of over 98% for microspheres, enabling rapid PSA detection. *Source*: [4] / IOP Publishing / CC BY 4.0.

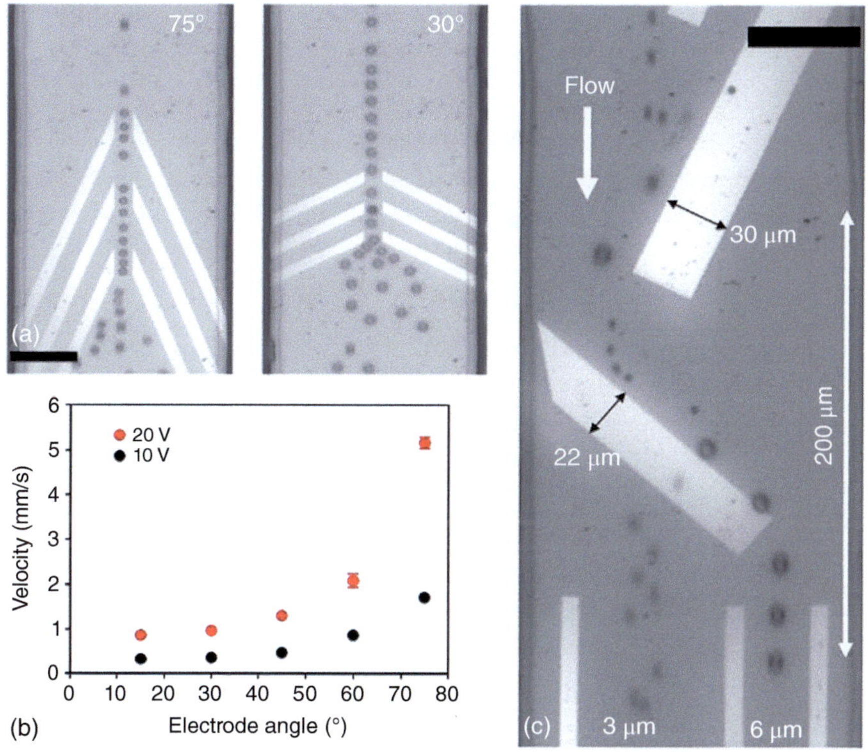

Figure 8.4 Integrated OET particle screening chip composed of an SU8 photoresist [4]. (a) Focus experiments of 6 µm particles with angled electrodes; (b) velocity of focused 6 µm particles depending on electrode angle and voltage signal; (c) guide and separate 6 and 3 µm beads into microchannels. *Source*: [4] / IOP Publishing / CC BY 4.0.

photolithography to form microchannels with defined geometries. These channels are enclosed between indium tin oxide (ITO)-coated glass slides, with one slide additionally coated with amorphous silicon to serve as the photoconductive layer.

The incorporation of OET into microfluidics offers notable benefits, such as the capacity to generate adaptable virtual electrodes through light patterns. This feature enables the dynamic and flexible control of particles within the fluidic channels, allowing for the manipulation, focusing, and sorting of microparticles with exceptional precision, even under continuous flow conditions. Experimental findings illustrate the device's capability to manipulate particles at speeds of up to 5 mm/s, underscoring its potential for applications in particle sorting and flow cytometry. The research underscores the reliable performance and versatility of the integrated OET-microfluidic platform, positioning it as a valuable asset for advanced microfluidic endeavors.

Lin and Lee introduced an innovative microfluidic chip in their research, combining OET with flow cytometry for continuous microparticle counting and sorting, as depicted in Figure 8.5a, b [5]. The chip fabrication process utilized standard micro-electro-mechanical-system (MEMS) techniques, where an ITO-coated glass

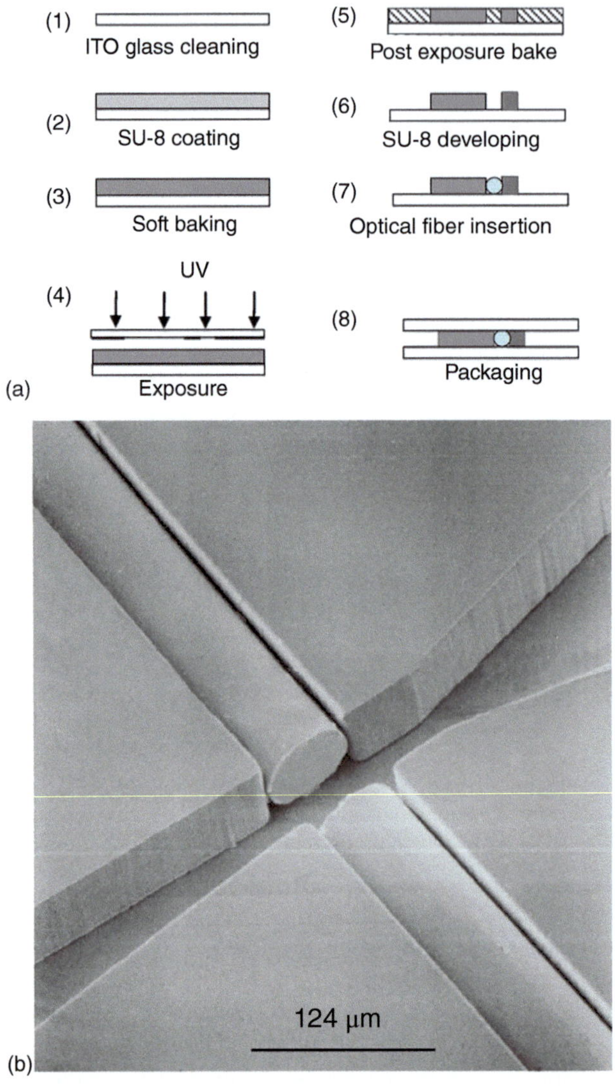

Figure 8.5 OET chip with integrated optical fiber [5]. (a) Simplified microfluidic chip fabrication; (b) SEM image of etched optical fiber in SU-8 structure. *Source*: [5] / with permission of Elsevier.

substrate was coated with an amorphous silicon to establish a photoconductive surface. A notable aspect of this design is the incorporation of buried optical fibers within the microfluidic channels, eliminating the necessity for intricate lithography or metal patterning.

The primary benefit of integrating ODEP with microfluidics lies in the ability to generate adaptable "virtual electrodes" using light, enabling dynamic particle manipulation without physical electrodes. This methodology effectively focuses, counts, and sorts microparticles based on size, utilizing the gravity-driven flow. Experimental findings showcased the chip's proficiency in sorting 20.9 and 9.7 μm

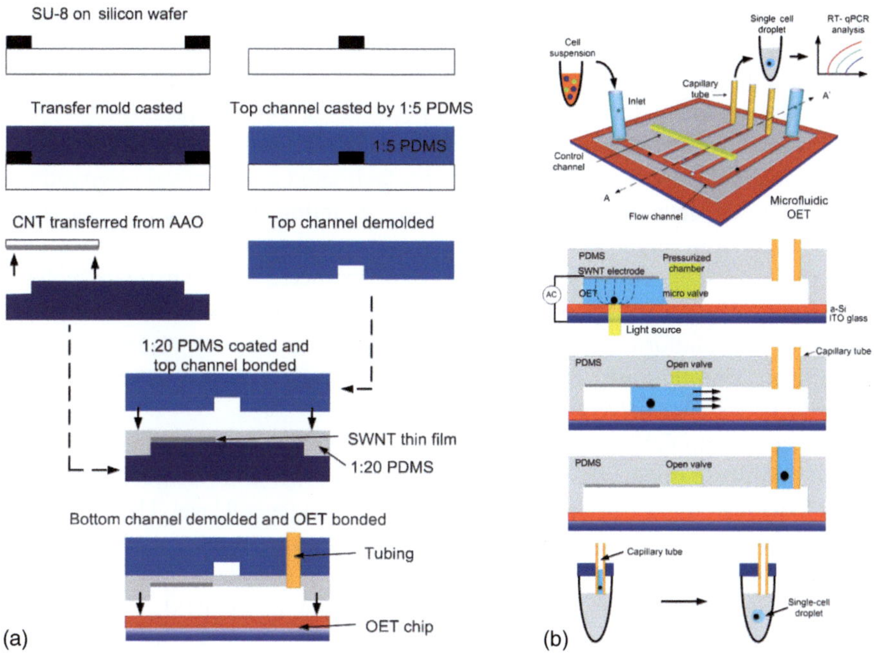

Figure 8.6 (a) Schematic of the OET-integrated microfluidic platform. (b) The procedure of preparing single-cell samples for analysis [6]. *Source*: [6] / The Royal Society of Chemistry.

polystyrene microparticles with precision, achieving sorting speeds of up to 11 particles per second for larger particles. The system's versatility and simplicity present a significant promise for diverse biomedical applications, including cell sorting and analysis.

Huang et al. developed a microfluidic platform integrated with OET specifically designed for single-cell preparation and analysis (Figure 8.6a, b) [6]. The fabrication of the platform involved embedding single-wall carbon nanotube (SWNT) electrodes within multilayer polydimethylsiloxane (PDMS) structures to create a transparent, flexible, and stable environment suitable for manipulating biological samples. The device comprises a bottom layer for fluid manipulation and a top layer with pneumatic valves to regulate fluid flow. The PDMS layers are bonded with a photoconductive OET chip through oxygen plasma treatment, establishing a durable and integrated system.

The construction of the microfluidic chip commences with the formation of SU-8 molds on silicon wafers, used to cast PDMS transfer molds. These molds aid in transferring the SWNT thin films onto the fluidic channels, which are essential for enabling OET manipulation within the microfluidic setting. The chip is assembled by bonding the multilayer PDMS device with the OET chip, ensuring precise alignment and operational efficacy.

The integration of OET with microfluidics offers numerous advantages. The OET system manipulates cells and particles within the microfluidic channels. This technique enables the selective, noninvasive trapping and movement of individual cells based on their optical characteristics, such as size, shape, and fluorescence. The capability to manipulate single cells in a massively parallel manner using light beams, coupled with the microfluidic system's fluid flow control, enhances the precision and throughput of single-cell analyses.

Experimental results showcased the platform's efficacy in isolating and manipulating individual cells. The SWNT electrodes facilitated clear optical observation and fluorescence detection within the device, enabling researchers to identify and selectively trap target cells. These cells were then transported into isolated chambers, encapsulated in nanoliter droplets, and extracted for downstream molecular analyses like reverse transcription quantitative polymerase chain reaction (RT-qPCR). The platform achieved a transport speed of around 10 μm/s for cell manipulation, slightly lower than traditional ITO-based OET systems due to the higher sheet resistance of SWNT electrodes.

The precise manipulation and analysis of single cells offered by this integrated system hold significant implications for biomedical diagnostics and research, providing a robust tool for studying cellular heterogeneity at the single-cell level. The combination of OET with advanced microfluidic control functions in this platform represents a notable advancement in lab-on-a-chip technology.

Yang et al. [7] introduced a groundbreaking Self-Locking OET (SLOT) platform that merges OET with the microfluidic technology to overcome the key limitations of conventional OET systems, particularly in high-conductivity media, as illustrated in Figure 8.7a, b. The microfluidic chip fabrication entails using a p-type single crystalline silicon substrate with a 1 μm SiO_2 layer patterned into an array of circular rings. NPN bipolar junction phototransistors are formed through phosphorus ion implantation, which are then coated with a thin Al_2O_3 dielectric layer to enable self-locking functionality in high-conductivity environments.

The SLOT system addresses challenges associated with manipulating single cells over large areas and in physiological buffers, where traditional OET systems encounter difficulties due to limited field-of-view (FOV) and photocurrent issues. The self-locking mechanism of SLOT relies on an array of phototransistor-controlled

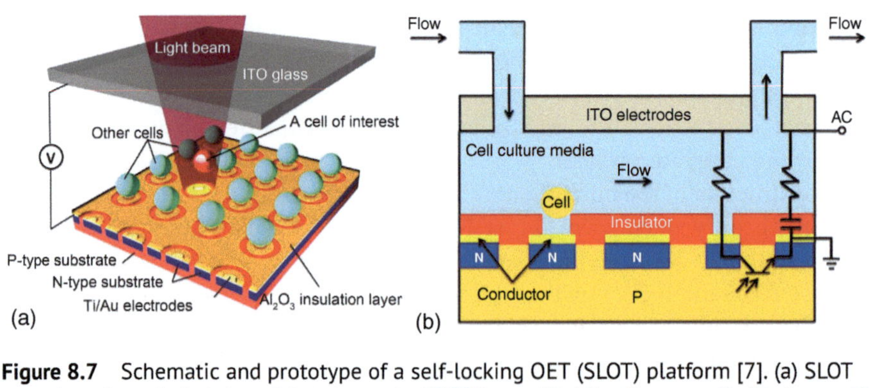

Figure 8.7 Schematic and prototype of a self-locking OET (SLOT) platform [7]. (a) SLOT uses ring-shaped phototransistors to control DEP traps; (b) AC voltage applied between ITO electrode and conductor layer, with current flowing laterally through transistors and Al_2O_3. Light turns off DEP, releasing trapped cells via fluid flow. *Source*: [7] / IOP Publications / CC BY 3.0.

traps that autonomously lock cells and particles without requiring continuous illumination. Upon targeted light exposure, the traps release the cells, which are subsequently transported by fluid flow.

Experimental findings showcased the SLOT platform's ability to trap over 100,000 microparticles across a $1\,cm^2$ area and selectively release them to form patterns. Furthermore, SLOT effectively manipulated single cells in standard cell culture media, maintaining high cell viability and normal multi-day divisibility post-manipulation. This platform marks a significant leap in large-area single-cell manipulation and holds broad applications in tissue engineering, drug screening, and cell sorting within the realm of biomedical research.

8.2 Photocuring Technology

Yang et al. [9] developed an innovative approach that combines OET with the photocuring technology to fabricate and manipulate microscale hydrogel structures for tissue engineering applications. Photocuring involves utilizing light, typically UV light, to initiate the polymerization of a photosensitive material, transforming liquid precursors into solid structures. The integration of OET with photocuring offers numerous advantages, including the ability to dynamically synthesize and reconfigure 3D microstructures with high precision and flexibility.

The photocuring process takes place within a microfluidic device using a digital micromirror device (DMD) to project UV light patterns onto a pre-polymer solution [8] (Figure 8.8a–h). The material chosen for photocuring is poly (ethylene glycol) diacrylate (PEGDA). Upon exposure to UV light, the PEGDA solution cross-links, forming solid microstructures suspended in the fluid, which are then transferred to the OET platform for further manipulation [10].

One of the primary challenges in this process is maintaining the uniformity and homogeneity of the hydrogel microstructures, particularly when dealing with complex shapes and varying sizes. This challenge was addressed by optimizing the DMD-based lithography system, enabling real-time control over the light patterns to ensure the precise fabrication of microstructures with well-defined

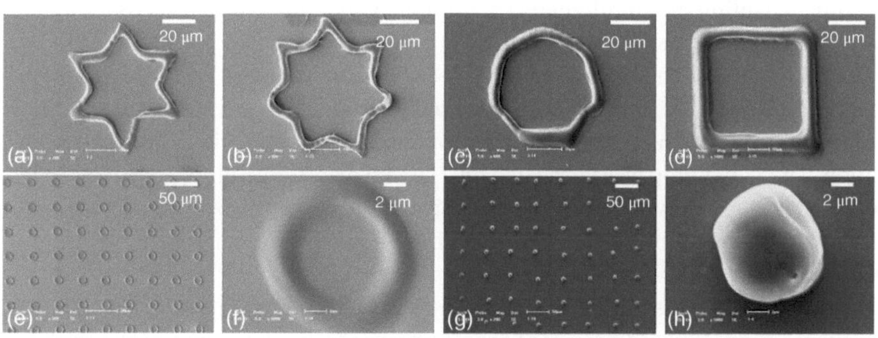

Figure 8.8 (a–d) show SEM images of polymer structures with different geometries. (e–h) show SEM images of different arrays of structures [8]. *Source*: [9] / MDPI / CC BY 4.0.

shapes and edges. Additionally, the study emphasizes the significance of controlling the photoinitiator concentration and UV exposure time to achieve consistent polymerization, thereby preventing the incomplete curing or excessive swelling of the hydrogel structures.

Experimental results exhibited the successful fabrication of various hydrogel shapes, such as triangles, stars, and customized geometries, with excellent structural fidelity. The OET platform effectively manipulated these structures, arranging them into complex 3D configurations. The capability to assemble microstructures containing different cell types into a single, reconfigurable unit was also demonstrated, showcasing the technology's potential for applications in tissue engineering, drug delivery, and regenerative medicine.

8.3 Electrowetting Technology

8.3.1 Digital Microfluidics

Digital microfluidics, particularly electrowetting on dielectric (EWOD), is a technique that enables precise control of droplets on a surface using electric fields, allowing for operations such as droplet movement, splitting, merging, and mixing. However, it faces limitations in handling and manipulating individual particles or cells inside droplets. By combining OET with EWOD, the optical control capabilities of OET can be leveraged to achieve the selective manipulation of individual particles or cells inside droplets. For instance, OET can accurately capture, move, and position specific cells through light-induced electric fields, a capability not achievable on traditional EWOD platforms.

One of the primary challenges in integrating EWOD and OET was the mutual exclusivity of their operational areas in the initial device design. In the first-generation device, the OET and EWOD regions were separate, restricting the ability to perform a continuous sequence of operations involving both technologies (Figure 8.9a) [11]. This challenge was addressed by developing a second-generation device that incorporates the lateral-field OET (LOET) technology (Figure 8.9b). This new design utilizes interdigitated electrodes on the same substrate, enabling the simultaneous operation of OET and EWOD across the entire chip without exclusive regions.

The fabrication of the integrated EWOD-LOET chip involved standard microfabrication processes. The EWOD chip was constructed on a glass substrate with ITO electrodes, coated with dielectric and hydrophobic layers to facilitate droplet manipulation. The LOET chip was fabricated on a silicon wafer, with aluminum electrodes patterned and coated with a photoconductive amorphous silicon layer. The combination of these technologies on a single chip allowed for precise manipulation of droplets and the cells within them.

Experimental results demonstrated the efficacy of the integrated device. The system successfully executed a series of operations where droplets containing live HeLa cells were manipulated using EWOD, followed by precise cell manipulation within the droplets using LOET, and finally, additional droplet operations using EWOD.

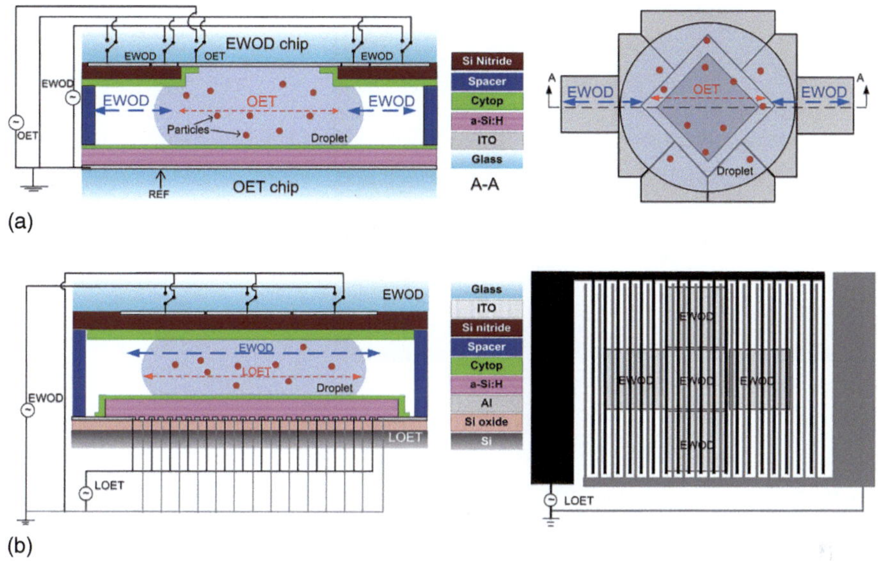

Figure 8.9 OET combined with electric wetting technology [11]. (a) The first generation combines EWOD and OET; (b) the second generation combines EWOD and new LOET. *Source*: [11] / The Royal Society of Chemistry.

The seamless performance of these operations on the same platform underscores the system's potential for complex cellular studies, enabling the isolation, manipulation, and analysis of individual cells within controlled microenvironments.

8.3.2 Optoelectrowetting Technology

Electrowetting technology often faces challenges related to the preparation of physical electrodes. Drawing inspiration from the OET technology, where light irradiation on optoelectronic materials can alter their properties, researchers have explored using light as virtual electrodes instead of physical ones. This approach significantly enhances the flexibility of electrowetting systems.

Park et al. introduced Single-Sided Continuous Optoelectrowetting (SCOEW), a novel method that enables the manipulation of droplets on a featureless photoconductive surface using light patterns [12] (Figure 8.10). Unlike traditional EWOD systems that rely on physical electrodes, SCOEW utilizes optical patterns to induce electrowetting, eliminating the constraints imposed by fixed electrode sizes. This technique allows for the continuous transportation, splitting, merging, and mixing of droplets across a wide range of volumes, from 50 mL to 250 pL, overcoming the size limitations of conventional EWOD platforms. The open, single-sided design of SCOEW also facilitates seamless integration with other microfluidic components, such as sample reservoirs, making it a versatile and scalable platform for lab-on-a-chip applications. An important advantage of SCOEW is its low-light intensity requirement, enabling droplet manipulation using common devices like liquid-crystal display (LCD) screens, underscoring its potential for portable and cost-effective microfluidic systems.

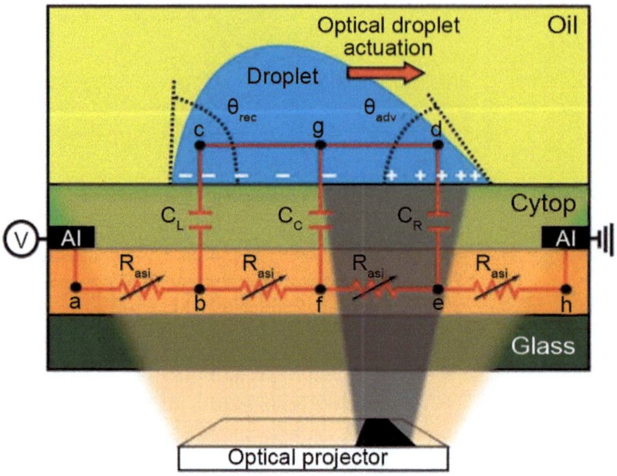

Figure 8.10 Schematics of the SCOEW device simply fabricated by a spin-coating method on a flexible polyethylene terephthalate (PET) substrate and its equivalent circuit [12]. *Source*: [12] / The Royal Society of Chemistry.

Furthermore, the challenge of operating droplets on flexible substrates has been addressed by optimizing the photoconductive layer and light patterning system to ensure consistent droplet control even when the substrate undergoes deformation. By projecting light patterns onto the flexible base, localized electric fields are induced to manipulate droplets in three dimensions, enabling complex operations such as vertical stacking and layer-by-layer assembly. This technique effectively maintains the precise droplet behavior even under bending or stretching conditions, showcasing the potential of optoelectrowetting in applications such as flexible electronics, soft robotics, and wearable devices, where traditional electrowetting methods may be less effective.

Despite the advancements in the optoelectrowetting technology, challenges related to droplet stability persist, particularly in open configurations or when droplets are subjected to external disturbances. In SCOEW, the study highlighted the occurrence of satellite droplets due to instability when large voltage differences are applied. To address this issue, the researchers optimized the thickness of the photoconductive layer and the dielectric properties of the surface to enhance droplet stability and control.

8.4 Image Recognition

8.4.1 Trajectory Planning

Trajectory planning is a crucial element in OET systems, especially when integrated with visual algorithms, as it enables the precise and efficient manipulation of microscale and nanoscale objects. A recent development in trajectory planning involves the utilization of synchronized OET-driven microrobots (OETdMs). This approach utilizes ODEP to control microrobots, allowing them to

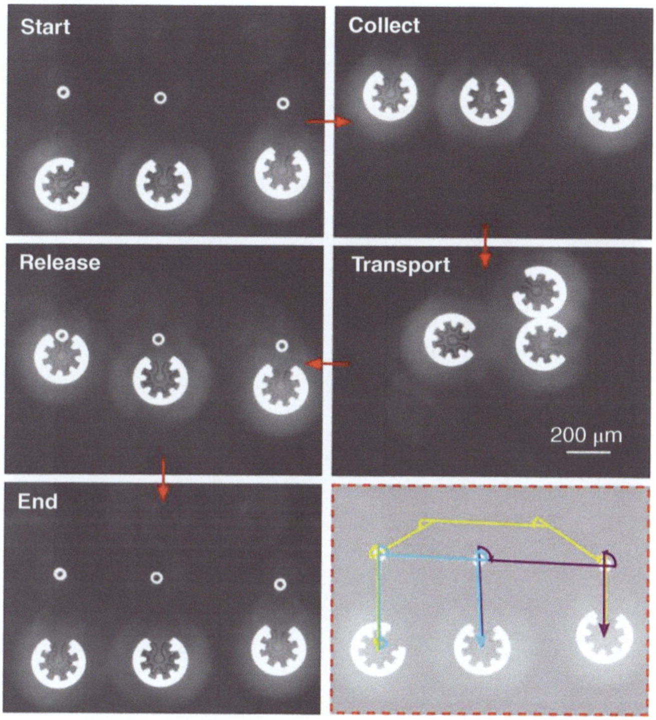

Figure 8.11 Microrobot and microsphere detection [13]. *Source*: [13] / IEEE.

autonomously perform complex micromanipulation tasks. The research demonstrates a technique for automated targeting and path planning, enabling the open-loop control of multiple microrobots simultaneously [13]. This method is particularly beneficial for tasks such as collecting, transporting, and depositing microscopic cargos like silica microspheres.

Utilizing low-level vision-based detection, the system identifies target objects, obstacles, and microrobots in the field of view. By framing the task as a capacitated vehicle routing problem (CVRP), the researchers optimized movement paths for multiple microrobots, allowing them to function autonomously in a complex, unstructured setting (Figure 8.11). The integration of a Multi-Agent Path Finding (MAPF) algorithm, specifically the Any-Angle Safe Interval Path Planning (AA-SIPP(m)) method, guarantees efficient and collision-free navigation for microrobots, even in crowded and cluttered environments. This study underscores the potential of employing OETdMs for high-throughput, parallel operations, significantly enhancing the trajectory planning capabilities of OET systems.

Liu et al. make a notable contribution to trajectory planning in OET systems, with a specific emphasis on precise micromanipulation. Their study introduces a real-time path planning strategy that utilizes POMDP (Partially Observable Markov Decision Process) to adjust to changing environments [14]. This technique enables the accurate manipulation of microscopic objects by constantly refining planned paths using real-time visual feedback (Figure 8.12a–c).

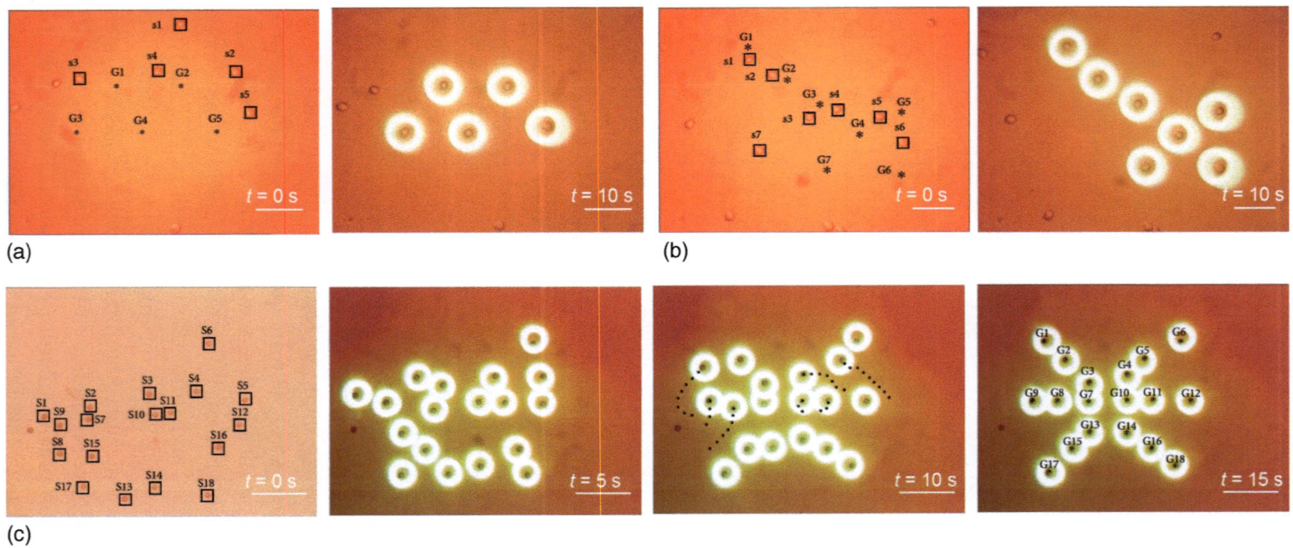

Figure 8.12 Experiments of applying the discretized manipulation method for micropatterning. (a) A cell navigated to its goal, avoiding obstacles; (b) two cells avoided colliding and reached their goals; (c) two cells and three polystyrene particles were transported to the goal positions. *Source*: [14] / American Association for the Advancement of Science - AAAS.

Tailored to tackle the challenges presented by the dynamic and unpredictable characteristics of microscale environments, such as Brownian motion and variable particle responses, the enhanced A-star algorithm optimizes the trajectory of manipulated particles. This approach guarantees efficient navigation around obstacles while upholding high precision [15]. The algorithm's efficacy is showcased through a range of microscale manipulation tasks, including the transportation of biological cells. The precise and adaptive control of the ODEP system notably enhances operational accuracy and reliability.

8.4.2 Measurement of Various Cell Parameters

The fusion of OET with image recognition technology has significantly propelled the measurement of cell mass, spectroscopic properties, velocity, and dielectric characteristics. In the realm of cell mass and density measurement, Zhao et al. showcased the utilization of an optically induced electrokinetics (OEK) microfluidic platform to accurately ascertain these parameters (Figure 8.13a–d) [16]. By leveraging computer vision algorithms to monitor the sedimentation trajectory of cells, the system swiftly gauges the volume, mass, and density of individual cells, providing a scalable solution for analyzing large quantities of cells simultaneously.

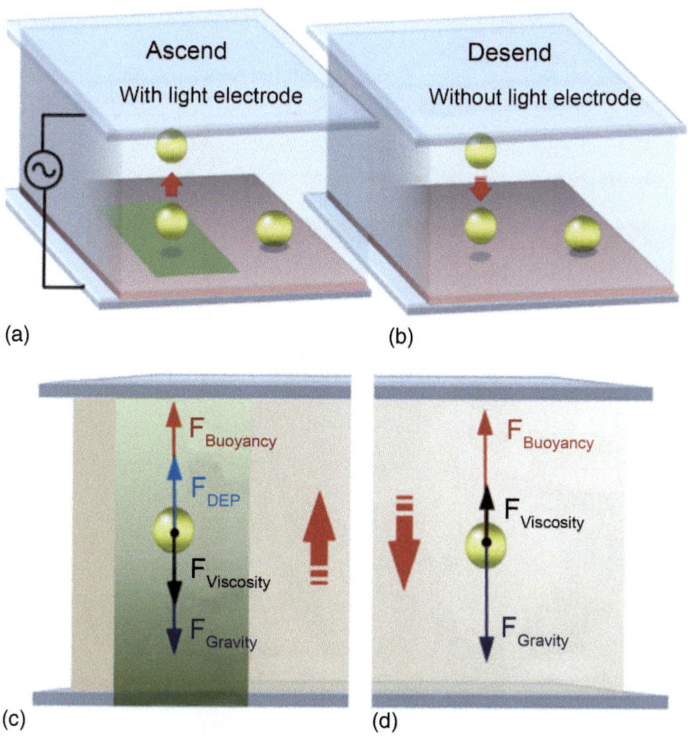

Figure 8.13 Using OET combined with image recognition technology to measure cell quality. (a) Light lifts the particle on the OEK chip; (b) without light, the particle sediments; (c) forces during rising; (d) forces during falling. *Source*: Adapted from [14].

In the context of spectral analysis, Zaman et al. introduced “spectral tweezers,” a pioneering approach that merges OET with spectroscopic analysis. This method utilizes a single focused light beam for both trapping force and spectral analysis, enabling the concurrent manipulation and real-time label-free characterization of individual micro-samples [17]. This technique proves particularly valuable for distinguishing distinct spectral signatures of single cells, facilitating the differentiation of cell types based on their spectral responses.

For velocity measurement, image recognition is harnessed to monitor the movement of cells manipulated by OET, offering a real-time feedback on their speed and trajectory [18] (Figure 8.14). The precision of these measurements is crucial for applications such as drug delivery, where the accurate control of particle movement is essential.

The dielectric properties of cells are evaluated through the integration of OET systems with image recognition. By examining the reaction of cells to different electric fields, it becomes feasible to deduce their dielectric characteristics, which offer insights into their physiological condition. This amalgamation of OET with

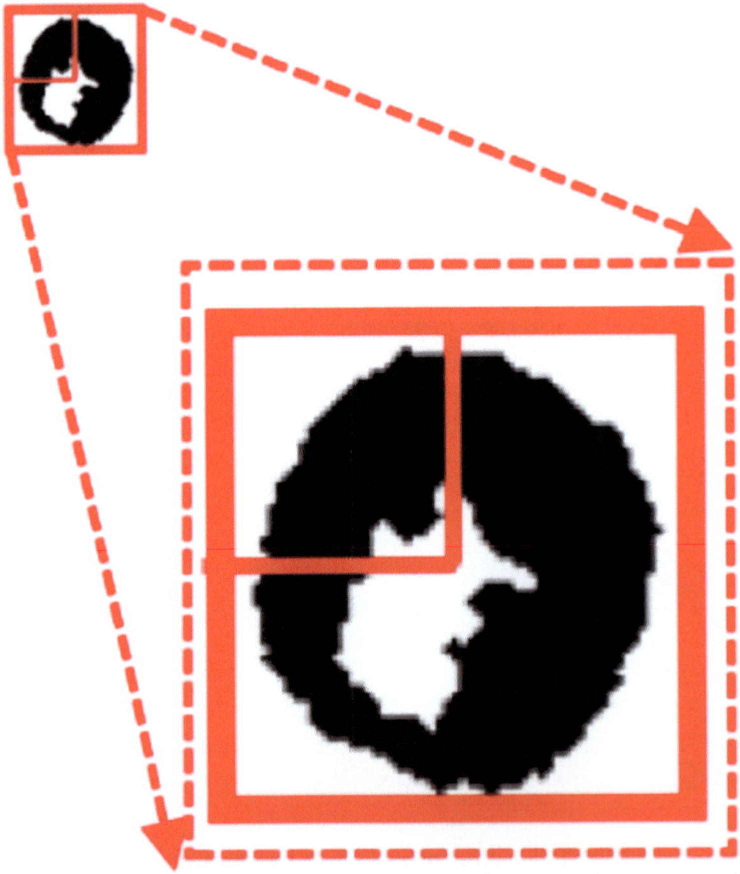

Figure 8.14 Measuring cell rotation speed using the visual recognition technology [17]. *Source*: [18] / MDPI / CC BY 4.0.

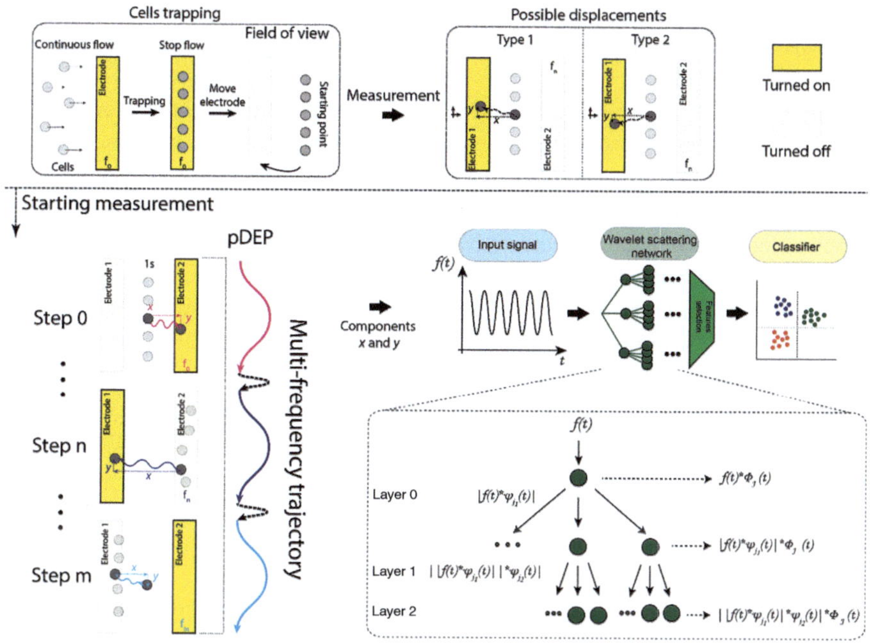

Figure 8.15 Cell electrokinetic fingerprint. *Source*: [19] / John Wiley & Sons.

advanced imaging methodologies has significantly heightened the accuracy and adaptability of cellular analysis, rendering it a potent instrument in both research and clinical diagnostics [19] (Figure 8.15).

References

1 Chen, Y.-S., Lai, C.P.-K., Chen, C., and Lee, G.-B. (2021). Isolation and recovery of extracellular vesicles using optically induced dielectrophoresis on an integrated microfluidic platform. *Lab on a Chip* 21 (8): 1475–1483.

2 Pei, S.N., Valley, J.K., Wang, Y.-L., and Wu, M.C. (2015). Distributed circuit model for multi-color light-actuated opto-electrowetting microfluidic device. *Journal of Lightwave Technology* 33 (16): 3486–3493.

3 Cheng, X.T. (2024). *Research on Fluorescence Microsphere Detection Technology Based on Microwell Array Optoelectronic Tweezers Chip*. Beijing, China: Beihang University.

4 C. Witte, J. Reboud, J. M. Cooper, and S. L. Neale, "Channel integrated optoelectronic tweezer chip for microfluidic particle manipulation," *Journal of Micromechanics and Microengineering*, vol. 30, no. 4, art no. 045004, 2020.

5 Lin, Y.H. and Lee, G.B. (2008). Optically induced flow cytometry for continuous microparticle counting and sorting. *Biosensors and Bioelectronics* 24 (4): 572–578.

6 Huang, K.W., Wu, Y.C., Lee, J.A., and Chiou, P.Y. (2013). Microfluidic integrated optoelectronic tweezers for single-cell preparation and analysis. *Lab on a Chip* 13 (18): 3721–3727.

7 Yang, Y., Mao, Y., Shin, K.S. et al. (2016). Self-locking optoelectronic tweezers for single-cell and microparticle manipulation across a large area in high conductivity media. *Scientific Reports* 6: 22630.

8 Wang, S., Liang, W., Dong, Z. et al. (2011). Fabrication of micrometer- and nanometer-scale polymer structures by visible light induced dielectrophoresis (DEP) force. *Micromachines* 2 (4): 431–442.

9 Yang, W., Yu, H., Li, G. et al. (2017). High-throughput fabrication and modular assembly of 3D heterogeneous microscale tissues. *Small* 13 (5): 1602769.

10 Yang, W., Wang, W., Teng, X. et al. (2024). Construction and manipulation of origami-inspired tubular structures with controlled mechanical buckling for collection and transportation of microspheres based on optically induced electrokinetics. *Science China Technological Sciences* 67 (9): 2954–2964.

11 Shah, G.J., Ohta, A.T., Chiou, E.P. et al. (2009). EWOD-driven droplet microfluidic device integrated with optoelectronic tweezers as an automated platform for cellular isolation and analysis. *Lab on a Chip* 9 (12): 1732–1739.

12 Park, S.Y., Teitell, M.A., and Chiou, E.P. (2010). Single-sided continuous optoelectrowetting (SCOEW) for droplet manipulation with light patterns. *Lab on a Chip* 10 (13): 1655–1661.

13 Bendkowski, C., Mennillo, L., Xu, T., et al. 2021. Autonomous object harvesting using synchronized optoelectronic microrobots. *IEEE/RSJ International Conference on Intelligent Robots and Systems (IROS)* (16 December), IEEE, pp. 7498–7504.

14 Liu, J., Wang, H., Liu, M. et al. (2022). POMDP-based real-time path planning for manipulation of multiple microparticles via optoelectronic tweezers. *Cyborg and Bionic Systems* 2022: 9890607.

15 Du, M., Liu, F., and Li, G. (2024). Precision manipulation by an optically induced dielectrophoresis system based on an improved A-star algorithm. *Applied Optics* 63 (20): 5237–5243.

16 Zhao, Y., Lai, H.S., Zhang, G. et al. (2014). Rapid determination of cell mass and density using digitally controlled electric field in a microfluidic chip. *Lab on a Chip* 14 (22): 4426–4434.

17 Zaman, M.A., Wu, M., Ren, W. et al. (2024). Spectral tweezers: single sample spectroscopy using optoelectronic tweezers. *Applied Physics Letters* 124 (7): 071104.

18 Wu, H., Dang, D., Yang, X. et al. (2022). Accurate and automatic extraction of cell self-rotation speed in an ODEP field using an area change algorithm. *Micromachines* 13 (6): 818.

19 Filippi, J., Casti, P., Antonelli, G. et al. (2024). Cell electrokinetic fingerprint: a novel approach based on optically induced dielectrophoresis (ODEP) for in-flow identification of single cells. *Small Methods* 8 (8): e2300923.

9

Case Studies and Experiments

The OET technology has garnered significant interest across academic and industrial sectors due to its unique effectiveness in diverse disciplines and practical applications. This chapter introduces a selection of experiments conducted using OET, providing an in-depth analysis of these studies. By presenting case studies from various applications, the chapter derives valuable insights and lessons from different scenarios.

9.1 In-Depth Analysis of Notable Experiments

9.1.1 Detection of Single B Cells Producing RBD-Neutralizing Antibodies

The rise of the SARS-CoV-2 Omicron variant poses a major global health challenge, primarily because of its resistance to numerous antibodies that were developed earlier in the pandemic. There is an urgent need for therapeutic antibodies that exhibit potent efficacy against the Omicron variant. In a study by Yu et al., memory B cells were enriched and activated into antibody-secreting cells over an 8-day period. These cells were then loaded onto the Beacon® single-cell optofluidic system OptoSelect® 11 k chip [1]. The culture medium, continuously perfused with fresh supplements to promote the plasma cell survival, allowed researchers to conduct multiple rounds of binding and functional screening on individual B cells (Figure 9.1a, b). Initially, the authors assessed whether immunoglobulin G (IgG) antibodies secreted by thousands of single B cells exhibited binding activity to the receptor-binding domain (RBD). This was achieved by linking biotinylated RBD to streptavidin-coupled polystyrene beads and loading them into the channels above the NanoPens. Individual B cells secreted antibodies into the NanoPens, and antibodies that bound to the RBD antigen were detected using fluorescent anti-IgG secondary antibodies. Reactive antibodies diffusing from the NanoPens were visualized as fluorescent "blooming." Through this approach, the researchers identified 137 RBD-binding antibodies from a total of 4246 single cells on the NanoPens.

To determine whether the RBD-positive cells could produce antibodies capable of blocking the hACE2 receptor's interaction with the RBD antigen, an additional

Optoelectronic Tweezers for Manipulation at the Micro and Nano Scale, First Edition.
Lin Feng, Menglu Tan, Ao Wang, and Jiaying Zhang.
© 2026 WILEY-VCH GmbH. All rights reserved, including rights for text and data mining and training of artificial intelligence technologies or similar technologies. Published 2026 by WILEY-VCH GmbH.

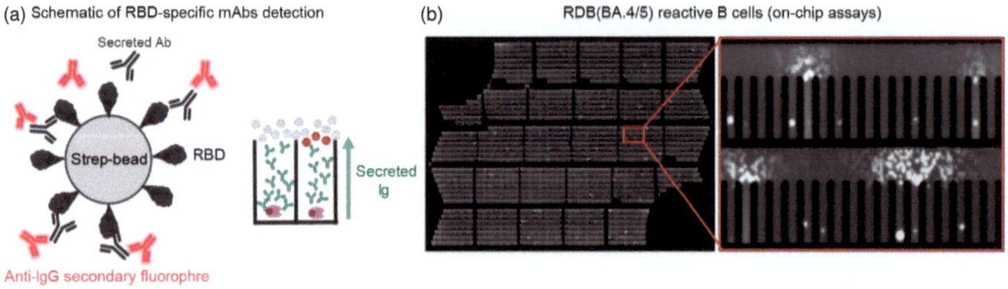

Figure 9.1 (a) Diagram of detecting RBD-specific mAbs. (b) Image of a full 11 k chip with a zoomed view highlighting "blooms" over pens with cells secreting RBD-reactive antibodies. *Source*: [1] / Frontiers Media S.A. / CC BY 4.0.

screening was conducted [1]. In this process, streptavidin microbeads linked with RBD were utilized, allowing them to bind directly to RBD-reactive antibodies for 30 minutes. Following this, soluble hACE2 was introduced to the chip to bind the RBD. Different fluorescence channels were employed to detect antibodies bound to RBD microbeads and the binding of hACE2 separately (Figure 9.2a, b). Interestingly, a single B cell in NanoPen 3462 secreted antibodies that interfered with hACE2 binding, shown by beads marked solely positive for RBD binding. Conversely, the B cell in NanoPen 9058 produced antibodies that connected to RBD but did not block hACE2, as evidenced by beads emitting positive signals for both.

In summary, Yu et al. [1] applied the Beacon® platform to extract monoclonal antibodies targeting the RBD of the SARS-CoV-2 S protein from individuals recovering from Omicron infections. Among their findings, two candidate antibodies,

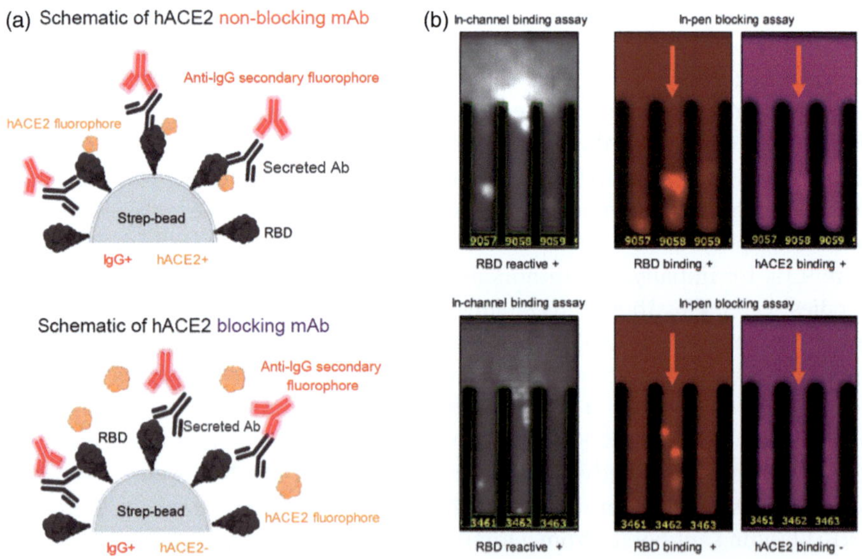

Figure 9.2 (a) Schematic illustrating the detection of secreted Ab and hACE2 binding on an RBD-coated streptavidin bead. (b) Identification of mAbs with hACE2-blocking activity using single-cell functional screening. *Source*: [1] / Frontiers Media S.A. / CC BY 4.0.

2173-A6 and 3462-A4, were notable for their strong RBD binding and robust neutralization capacity against various Omicron-related pseudoviruses. The efficiency of the Beacon® platform facilitated the hurried extraction and pinpointing of potential neutralizing antibodies from human samples within just 10 days. This streamlined method in antibody discovery is anticipated to simplify and accelerate future workflows, paving the way for the rapid development of new therapeutic antibodies. These advancements will be crucial in dealing with emerging SARS-CoV-2 strains and potentially combating future unknown infectious diseases.

9.1.2 Assuring Clonality of Single Cell Development

In the field of biopharmaceuticals, cell line development is a crucial step in the production of biotherapeutic drugs. Regulatory agencies require that all cell lines used for biomanufacturing must be derived from a single-cell progenitor to ensure consistency in safety and product quality. Traditional single-cell cloning methods suffer from resource-intensive and time-consuming issues, while emerging digital cell culture technologies, such as the Beacon platform, hold promise in addressing these bottlenecks. This study aims to validate the single-cell cloning capability of the Beacon platform in cell line development, particularly in ensuring efficiency and reliability in clonality. By comparing with traditional methods, the research [2] seeks to demonstrate the advantages of the Beacon platform in single-cell cloning processes, including higher clonality assurance, reduced resource consumption, and a more efficient cell line development workflow. This research is of significant practical importance to the biopharmaceutical industry as it can directly impact the development efficiency and quality control of biotherapeutic drugs.

The research methods employed by Le et al. [2] included limiting dilution, fluorescence-activated cell sorting (FACS), and cloning using the Beacon platform (Figure 9.3a, b). Limiting dilution involved diluting cells to approximately 0.7 cells per well based on the cell density and seeding them into a 96-well plate. Fluorescence-activated cell sorting utilized the FACS AriaIIu cell sorter to directly sort single cells into a pre-filled culture medium in a 96-well plate, with single-cell deposition verified under a microscope. Cloning with the Beacon platform involved the use of the Beacon instrument's optical manipulation technology to load single cells into NanoPens on a nanofluidic chip, with real-time imaging and cell counting facilitated by integrated microscopy. Throughout the cloning process, a combination of microscopy techniques was employed to ensure single-cell deposition, culture isolation, and final cloning purity. The Beacon platform specifically incorporated a chip tilting design and laminar flow control to prevent cross-contamination between NanoPens. These methods enabled the researchers to conduct single-cell cloning experiments on transfected Chinese hamster ovary cells, providing a reliable experimental foundation for subsequent studies.

The research results (Figure 9.4a–d) demonstrate that the Beacon platform exhibits extremely high accuracy and purity in the single-cell loading, NanoPen isolation, and cell export stages, with purity levels reaching 99.7%,

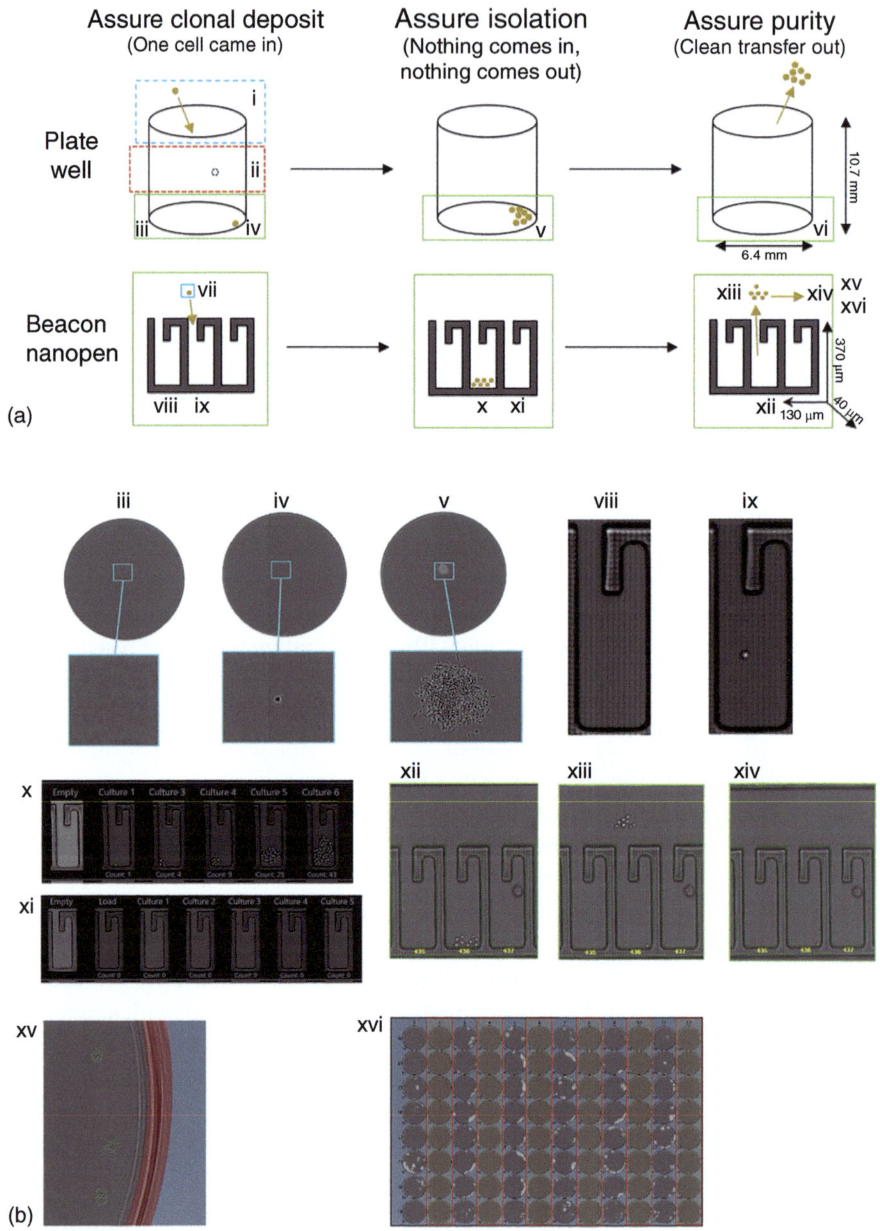

Figure 9.3 Workflow for subcloning and controls for single-cell progenitors. (a) Schematic of comparing single cell cloning workflows associated with microtiter plate-based cloning (top) and Beacon platform (bottom); (b) examples of image evidence used to support single cell progenitor assurance. *Source*: [2] / John Wiley & Sons / CC BY 4.0.

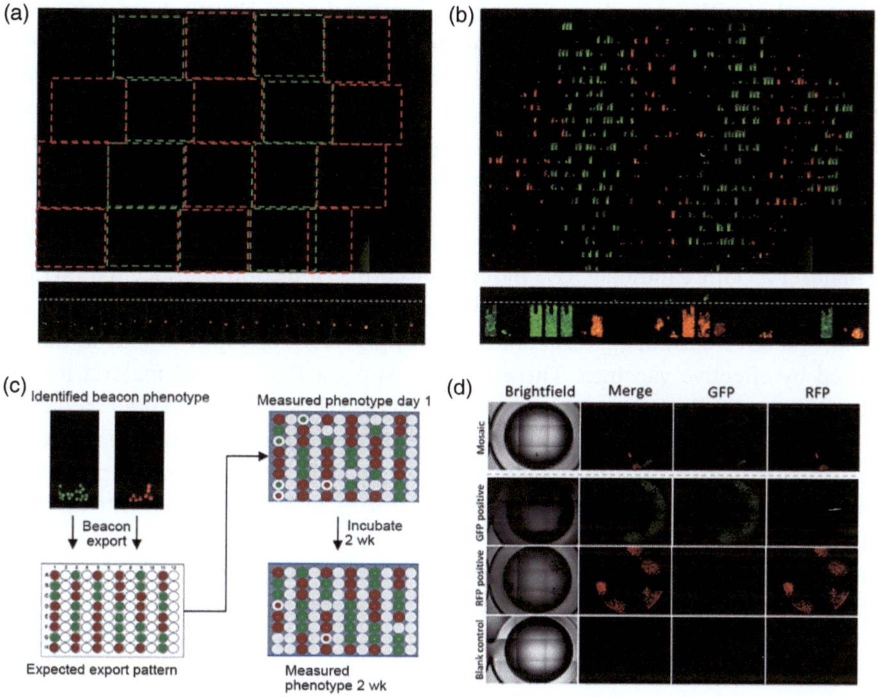

Figure 9.4 Validation of workflow stages utilizing a representative model CHO system. (a) Single cells expressing GFP or RFP were loaded onto specific chip zones, with red and green fluorescence merged in images; (b) cells are cultured to ensure no mixing between pens, marked by a white dotted line; (c) exported cells with expected phenotypes are confirmed by imaging well plates using a fluorescent imager; (d) images of fluorescent protein clones in well plates show mixed colonies acquired and scaled independently. *Source*: [2] / John Wiley & Sons / CC BY 4.0.

100%, and 100%, respectively. Compared to FACS and limiting dilution methods, the Beacon platform shows a significant advantage in terms of cloning purity assurance, with rates of 100%, 98.52%, and 96.08%, respectively. Additionally, the cloning efficiency of the Beacon platform is notably higher than that of other methods, at 52%, compared to 11% for FACS and 5.7% for limiting dilution. In terms of cell recovery rates, the Beacon platform demonstrates a high efficiency of 56%, far surpassing FACS at 24% and limiting dilution at 33%. Long-term validation results indicate that after 10 days of extended cultivation, the Beacon platform did not observe any cross-contamination events, further confirming its outstanding performance in preventing cell cross-contamination. These findings suggest that the Beacon platform offers greater flexibility and efficiency in cell selection and cultivation processes, providing a reliable technical support for single-cell cloning experiments.

9.1.3 Isolation of RH5-Specific B Cells

Malaria, a deadly disease caused by Plasmodium parasites, has posed a significant challenge to global public health. Despite the availability of some vaccines and drugs against malaria, the complex life cycle and variability of Plasmodium parasites continue to make prevention and treatment difficult. Developing vaccines targeting the blood stage of the parasite has been particularly challenging due to the variability and crypticity of the key antigen RH5 during the infection of red blood cells. Therefore, studying the antibody response against RH5 induced by natural infection and vaccination in the human body is crucial for designing more effective malaria vaccines.

Wang et al. conducted a comprehensive analysis of the antibody response against the malaria vaccine candidate RH5 induced by natural infection and vaccination [3]. Their study revealed that natural malaria infection can stimulate the production of rare but potent neutralizing antibodies targeting protein epitopes similar to those induced by effective vaccines. These findings suggest that natural malaria infection may enhance the immune response elicited by RH5 vaccines and offer valuable insights for the development of the next generation of RH5 vaccines. The Beacon® single-cell optogenetic system was utilized to screen 186 RH5 monoclonal antibodies from peripheral blood mononuclear cells (PBMCs) of patients infected with malaria and vaccinated individuals.

Wang et al. aimed to conduct a detailed analysis of the quality of the antibody response by isolating RH5-specific antibodies. PBMCs from individuals with high serum RH5 IgG levels were selected for comparison of antibody differences between natural infection and vaccination [3]. Additionally, PBMCs from individuals vaccinated with the RH5 vaccine but not infected were collected. Memory B cells were activated in 384 or 96-well plates using feeder cells and Staphylococcal enterotoxin B to compare antibodies produced from natural infection and vaccination. Antibody-secreting cells in the positive cell population were screened using the Beacon® single-cell optogenetic system. Researchers successfully isolated 164 RH5-specific IgG antibodies from vaccinated individuals (screening 135,000 IgG + B cells), while only 22 antibodies were isolated from individuals with natural infection (screening 14.4 million IgG + B cells). Growth inhibition assay (GIA) analysis of these 186 antibodies revealed that most antibodies isolated from individuals with natural infection lacked neutralizing ability, but two antibodies exhibited potent neutralizing capabilities (Figure 9.5a–h).

The Beacon® single-cell optogenetic system plays a crucial role in efficiently screening and identifying RH5-specific monoclonal antibodies from individuals with natural infection and vaccination. The evaluation and comparison of antibodies produced after natural infection and vaccination hold significant importance in guiding the subsequent optimization of vaccine design and enhancing vaccine efficacy for malaria. Here are some key reasons why this analysis is essential:

(1) *Identification of protective antibodies*: By analyzing antibodies from both infected patients and vaccine recipients, researchers can pinpoint effective and protective neutralizing antibodies. This analysis helps in identifying critical antigenic epitopes, which serve as valuable targets for vaccine design.
(2) *Guiding vaccine design*: Understanding the characteristics of these antibodies, such as the specific antigenic epitopes they target and their neutralization mechanisms, can provide valuable insights for scientists in designing

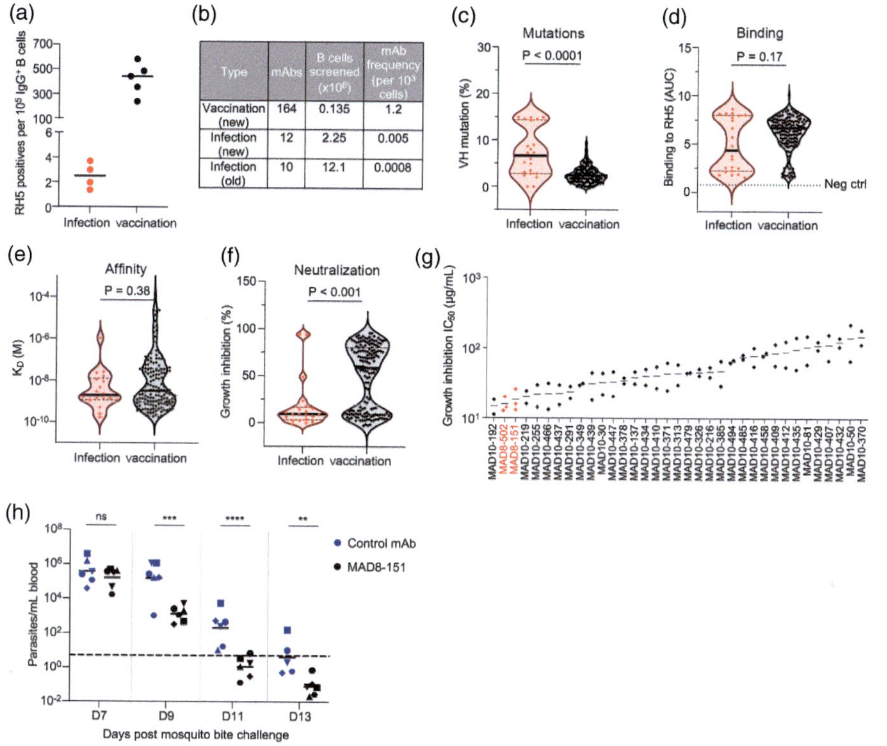

Figure 9.5 Malaria exposure elicits rare but potent neutralizing antibodies targeting RH5. (a) RH5-positive wells frequency in cultured IgG+ B cells from infected vs. vaccinated donors; (b) number of mAbs isolated from both donor types using a "new" sensitive and an "old" method; (c) variable heavy chain mutation mutation frequencies in mAbs from infected vs. vaccinated donors; (d) RH5 binding of mAbs; (e) binding affinity of mAbs from both donor types; (f) growth inhibition of mAbs isolated from infected versus vaccinated donors; (g) IC50 values for 35 potent mAbs, with each data point representing an experiment and bars showing mean values; (h) parasitemia in FRG HuHep mice challenged by infectious mosquito bite. *Source*: Adapted from Jennitte, et al., 2019.

vaccines that can elicit similar protective antibody responses. This knowledge helps in developing more effective and targeted vaccines.

(3) *Evaluating vaccine efficacy*: By comparing the antibody repertoires generated after natural infection and vaccination, researchers can assess whether the vaccine is capable of inducing immune responses that are comparable to or even superior to those elicited by natural infection. This evaluation aids in determining the potential effectiveness of the vaccine and serves as a foundation for further refinement and enhancement of the vaccine.

In conclusion, the analysis and comparison of antibodies produced following natural infection and vaccination using advanced technologies like the Beacon® single-cell optogenetic system offer valuable insights that can drive the development of more effective malaria vaccines and ultimately contribute to the global efforts in combating this deadly disease.

9.1.4 Screening of Antibody-Secreting Single B Cells

The persistent presence of leukemia stem cells (LSCs) is considered a major factor contributing to the treatment failure in patients with acute myeloid leukemia (AML). LSCs possess the ability for unlimited self-renewal and demonstrate high resistance to therapy, making them challenging to target and eliminate. One promising target for the AML treatment is the Interleukin-1 receptor accessory protein (IL1RAP), which is highly expressed on AML cells and LSCs but not on normal hematopoietic stem cells (HSCs). This specificity provides a new opportunity for selectively targeting and eliminating AML cells while preserving normal hematopoietic function.

In a study by Zhang et al. [4], Balb/c mice were immunized with the recombinant extracellular domain (ECD) of human IL1RAP to develop antibodies against human IL1RAP for leukemia immunotherapy (Figure 9.6a–g). CD138+ B cells from the spleens of immunized mice were screened using the Beacon® single-cell optogenetic system to identify single B cells capable of secreting IL1RAP-specific antibodies. Through single-cell RT-PCR, 34 pairs of VH/VL sequences of antibodies against human IL1RAP were isolated, and 12 monoclonal antibody (mAb) candidates were selected based on the sequence diversity of the complementarity-determining regions (CDRs). These 12 VH/VL sequences were then used to create mouse–human chimeric anti-IL1RAP antibodies by cloning them into vectors containing human constant domains. Subsequent characterization analyses led to the identification of IL1RAP-24 mAbs as the most potent and stable in terms of activity.

Using the Fab arm exchange technique (FAE), IL1RAP-24 mAbs and a humanized CD3 antibody targeting T cells were converted into IL1RAP × CD3 TCE. By introducing mutations in the Fc region of the antibodies to eliminate non-specific binding to FcγRs, the researchers obtained the BIF002 TCE antibody. Further mutations were introduced in the IL1RAP and CD3 Fab regions of BIF002 to design the CD3-arm-inactivated BIF018 (retaining IL1RAP recognition but lacking T-cell redirecting capability) and the IL1RAP and CD3 dual-arm-inactivated BIF026 (simultaneously lacking IL1RAP recognition and T-cell redirecting capability) as negative control antibodies. This innovative approach holds promise for the development of targeted immunotherapies for AML by selectively eliminating AML cells while sparing normal hematopoietic function.

Monoclonal antibodies not only act as standalone medications but also serve as the foundation for modifying functionally specific antibodies to create novel bispecific antibodies, antibody–drug conjugates, antibody fusion proteins, and other therapeutic options. The Beacon® single-cell optogenetic system offers a convenient and effective tool for researchers to identify monoclonal antibodies that target new antigens and develop innovative antibody therapies, which in turn supports the advancement of new drugs and breakthroughs in the field.

9.1.5 Development of CAR Cell Library and CAR Screening

The advancement of the chimeric antigen receptor T-cell therapy (CAR-T therapy) hinges on the exploration of targets and the subsequent screening of

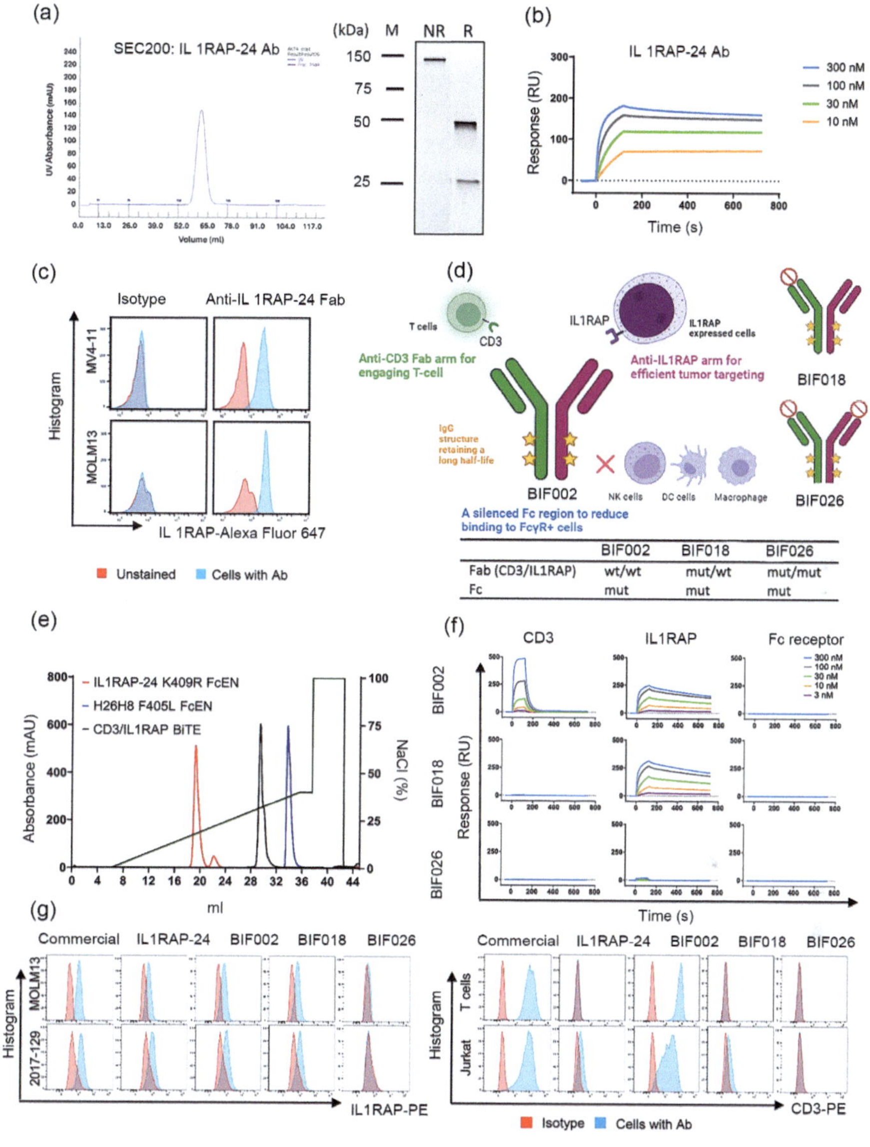

	BIF002	BIF018	BIF026
Fab (CD3/IL1RAP)	wt/wt	mut/wt	mut/mut
Fc	mut	mut	mut

Figure 9.6 Characterization of the anti-IL1RAP-24 monoclonal antibody (mAb) and production of the anti-IL1RAP/CD3 bispecific T-cell engager (TCE). (a) Size exclusion chromatography (SEC, left) and SDS-PAGE (right) of anti-IL1RAP-24 mAb; (b) target affinity assay of anti-IL1RAP-24 mAb binding to IL1RAP by surface plasmon resonance (SPR); (c) flow cytometry (FCM) analysis of binding capacity of anti-IL1RAP-24 Fab vs. isotype control antibody to IL1RAP^{pos} MV4-11 and MOLM13; (d) schematic description of anti-IL1RAP/CD3 bispecific TCEs including BIF002, BIF018 and BIF026; (e) Ion exchange chromatography (IEC) of anti-IL1RAP/CD3 bispecific TCE (BIF002) along with parental antibodies; (f) SPR sensorgrams of TCEs; (g) FCM analysis of binding capacity of anti-IL1RAP or anti-CD3 commercial mAbs, anti-IL1RAP-24 mAb, and TCEs including BIF002, BIF018 and BIF026, to IL1RAPhigh AML cells and CD3+ T and Jurkat cells. *Source:* Adapted from Lawrence, et al., 2024.

functional CARs. However, current cell therapies are limited to a small range of common targets [5–7]. It is imperative to identify additional targeting patterns to broaden the scope of cell therapy applications. Tailoring personalized cell therapies to individual cancer patients could be a pivotal strategy in enhancing patient outcomes in the future [8]. Therefore, the swift screening of effective CAR sequences will play a vital role in the progression of these therapies.

Conventional methods for the CAR discovery involve analyzing the cytokine release or cytotoxicity at a population level, necessitating multiple iterations to identify viable sequences. These methods lack single-cell resolution to directly pinpoint targets [9]. Tetramer staining techniques can only detect CARs that bind to targets without offering direct functional insights on cytotoxicity or cytokine secretion [10], leading to cumbersome re-expression and validation processes. The Beacon® single-cell functional characterization allows researchers to directly evaluate the phenotype and function of individual T cells interacting with antigen-presenting cells/beads in NanoPen® chambers, facilitating the retrieval of target T cells for CAR sequencing. The "function-first" screening strategy of Beacon® offers an efficient approach for researchers to discover CARs targeting diverse antigens, significantly reducing the time required compared to traditional techniques, from weeks or months to just a few days [11].

Schmidt et al. concentrated on streamlining the development process of CAR-T cells by tackling the obstacles encountered in the CAR discovery and clinical manufacturing. Ran and Schmidt incorporated the CliniMACS Prodigy system into their CAR engineering process and sought a rapid method for identifying functional CARs [12]. By utilizing the Beacon® platform's capacity to clone and functionally characterize thousands of single cells, they established the world's first CAR cell library and employed it for the screening of functional CARs.

To create host cell lines for constructing the CAR cell library, Schmidt et al. engineered Jurkat cells with a 12 × NFAT promoter driving the expression of the Venus GFP fluorescent reporter protein. The 12 × NFAT promoter can induce the expression of the Venus fluorescent protein and the intracellularly fused CAR structure (~260 aa). This design allows researchers to utilize fluorescence as an indicator for stable transduction. Subsequently, stable Venus GFP-expressing proliferative clones were identified and retrieved using the Beacon® platform.

A CAR cell library was generated from Jurkat cells that expressed Venus protein using non-viral DNA vectors. Cells with tonic signaling were initially enriched through FACS, then loaded onto the OptoSelect chip and co-cultured with antigen-presenting microbeads. The recognition of antigen-specific cells and subsequent cell activation led to the transcription of 12 × NFAT. Cells that expressed Venus GFP were singled out, transferred from the chip to a 96-well plate, and sequenced to pinpoint functional CARs. Following a 24-hour co-culture period, the heightened expression of Venus-GFP in Jurkat cells expressing CAR indicated successful cell activation, with 30 hits of varying specificities identified. These hits demonstrated the recognition of the target antigen on the bead surface and effective activation. Subsequently, 12 Jurkat cells expressing CAR were successfully isolated and sequenced, each cell expressing distinct CARs.

For the successful development and translation of the CAR-T therapy, a thorough comprehension of the cellular behavior is essential due to its nature as a cell-based complex product. The Beacon® platform stands out as a robust single-cell functional characterization technology capable of simultaneously assessing cell cytotoxicity, cytokine secretion, surface markers, and more. The detailed examination of single-cell function offers valuable insights into targeting patterns, cellular product heterogeneity, donor variability, and other critical aspects throughout the stages of the CAR discovery, validation, and clinical translation. Additionally, the Beacon® platform allows for the export of cells post data collection, unveiling the genomic and transcriptomic mechanisms that govern the cell behavior. This capability enables the integration of multi-omics workflows and facilitates a comprehensive understanding of the CAR-T therapy.

9.2 Case Studies Showcasing Diverse Applications

Gan et al. conducted a comprehensive study on the force exerted on microparticles in an OET system [13]. They utilized the theory of polarization and Maxwell stress tensors to explore the mechanisms of dielectrophoretic forces. By employing commercial arbitrary Lagrangian–Eulerian finite element software, they simulated the coordinated movement of microparticles in a fluid, solid, and electric field environment. A frequency-domain numerical model was developed to analyze the dielectrophoretic forces within the OET chip, aiding in the optimization of power and medium selection. The use of simulation techniques has offered valuable insights for the development of a more efficient OET system and has facilitated operations under diverse experimental conditions.

Through a feedback control strategy, the precise manipulation of microparticles has been achieved, including tasks such as image correction, target detection, and proportional-integral-derivative (PID) control. The control method incorporating the position feedback has significantly improved the control accuracy, with the experiment successfully maintaining the positional error of 15 μm diameter polystyrene microspheres within 15 μm, thereby enhancing operational capabilities in complex environments. Furthermore, by projecting images through the OET system, large-scale microparticle pattern assembly has been accomplished, with specific patterns created within a minute. By utilizing light patterns of varying brightness to create a light funnel, continuous size-based sorting of 8 and 15 μm polystyrene microspheres was successfully carried out, achieving a sorting efficiency of 81.89%.

In addition to their previous work, Gan et al. have developed a monolithic optoelectrofluidic chip by employing laser micromachining to fabricate electrodes in the form of micro-grooves, integrated with a microfluidic system [14]. This chip is designed to create a transverse electric field in its front segment by combining individual grooves and electrical signals, utilizing a dynamic photonic funneling pattern to concentrate macrophages. Addressing a common limitation in existing optical tweezer systems, where the AC electric field induced by projected light lacks rotational capability, the chip's rear segment features a cross-groove structure synchronized with electrical signals that are 90° out of phase. This configuration, when combined with a specific light pattern projection, generates a rotating electric field, effectively rotating microparticles. This method of AC rotation enables the characterization of the electrical properties of cells [15].

On the same chip, through a combination of various groove designs, superposition of electrical signals, and projection of light patterns, both electrofocusing and electrorotation effects are achieved simultaneously. This enables the characterization of the dielectric properties of macrophages, providing a comprehensive understanding of the cell behavior and properties through optoelectrofluidic manipulation.

In a foundational application within the OET system, Liang et al. focused on manipulating living cells using the unique dielectric properties of particles to achieve capture, transport, and release of micro-objects [16]. They conducted a detailed analysis of the polarization charge distribution and electric polarization vector distribution within homogeneous objects in the OET system to understand how dielectric particles respond to positive or negative dielectrophoretic forces. Proximity to the light pattern led to a higher density of polarization charges around the particles compared to those located further away. Simulation results revealed that the nonuniform polarization induced by the electric field's nonuniformity is akin to multipole moments within the particle.

In their methodology for cell transportation, Liang et al. exploited the attractive forces between particles with opposite dielectric properties to enable adhesion to the target particle, achieving an adhesion ratio of 1:5 while maintaining a velocity of 40 μm/s. This interaction was harnessed to enhance transport throughput and assemble specific microstructures using particles. The researchers selected motile diatoms and Spirulina with contrasting dielectric properties to interact with particles (Figure 9.7a, b). These microorganisms served as carriers for transporting polystyrene microbeads to designated locations and releasing them upon exiting the electric field. Micro-objects of various dimensions and topological structures, such as spherical polystyrene microspheres, rod-shaped diatom microspheres, and helical Spirulina microspirals, could interact within the optoelectrostatic trap system.

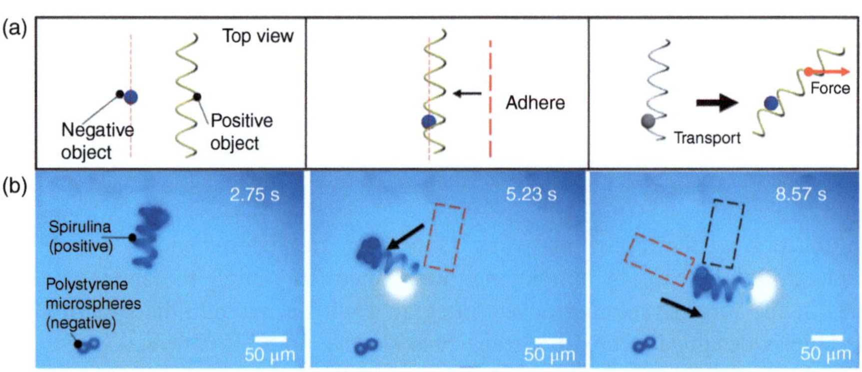

Figure 9.7 (a) Diagram illustrating the interaction between polystyrene microbeads and micro-Spirulina. (b) Movement of polystyrene microbeads along with Spirulina, where rectangles and black arrows indicate the previous positions and the direction of movement. *Source*: [16] / The Royal Society of Chemistry.

Importantly, these microorganisms remained unharmed during the experiments, and the interactions between particles demonstrated potential applications in the medical field, including cancer or virus detection, drug delivery, and a broad range of similar tasks. This innovative approach showcases the versatility and potential of OET manipulation in various biological and medical applications.

In their study, Liang et al. explored the manipulation of Spirulina within an OET system to facilitate the assembly of complex-shaped flexible structures, demonstrating the system's potential for diverse applications [17]. Initially, the researchers characterized the manipulation forces acting on Spirulina with intricate geometries using the Maxwell stress tensor method. Through simulations, they observed that micro-Spirulina within the optoelectrostatic trap system were primarily attracted to the light pattern due to positive dielectrophoretic forces.

By harnessing positive photoelectric-induced DEP and coordinating multiple light regions, Liang et al. successfully transported the micro-Spirulina chains of varying lengths, composed of different quantities of micro-Spirulina. They achieved micro-Spirulina chains with a size of up to 806.45 μm by linking the chains of varying lengths. The researchers found that the transport velocity of the chains was influenced by their length, with drag and gravitational forces affecting the movement of excessively long chains. Various operational modalities were explored, showing that micro-Spirulina could be translated in parallel by multiple light points in the same or opposite directions and could rotate in conjunction with specific points on the bottom electrode or on its own.

Utilizing the optoelectrostatic trap, Liang et al. successfully assembled micro-Spirulina into design configurations such as T-shaped circuits, connecting rods, and microcoil pairs (Figure 9.8a–c). This approach allowed for the accommodation of specific structural requirements and integration with other shaped micro-objects for collective assembly. The research suggests that assembly methodologies using a simple optoelectrostatic trap platform could find applications in smart materials for medical and microdevice applications, including soft sensors, microcircuits, and microsprings. This innovative work highlights the versatility and potential of OET systems in the assembly of complex and flexible microstructures.

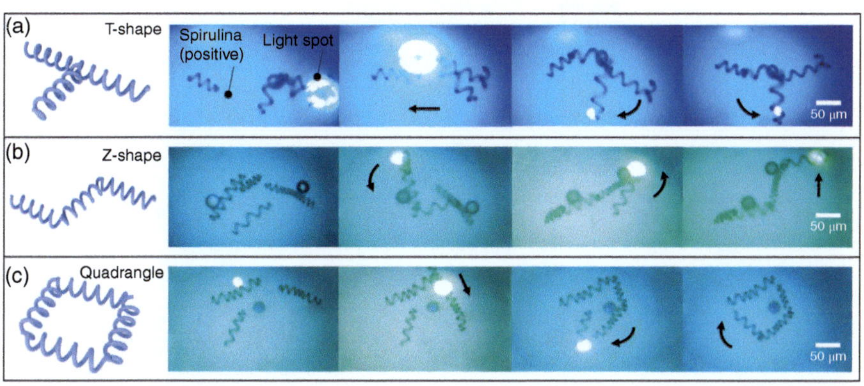

Figure 9.8 Flexible assembling microspiral into the different shapes of microstructure: (a) T-shape, (b) Z-shape, and (c) quadrangle shape. *Source*: [4] / Springer Nature / CC BY NC-ND 4.0.

The integration of OET systems with microfluidic devices has significantly expanded the application possibilities of the OET technology. Microfluidic devices enable the precise control and manipulation of fluids within small channels, typically submillimeter in size, under geometric constraints. By integrating microfluidic channels with OET systems, researchers can ensure an optimal thickness of the solution medium layer in the OET, facilitating continuous transportation and extraction of samples to the operational area of the OET. This integration enables high-throughput sorting, separation, and processing of target entities based on OET principles.

Huang et al. developed a vertically oriented photoelectrically induced DEP chip by etching channels on a double-sided adhesive tape, creating a microfluidic device for cell manipulation (Figure 9.9a) [18]. Within this device, dielectrophoretic forces were applied between two parallel laminar flows based on differences in cell size and electrical properties, allowing for the separation of cancer cells from a background of white blood cells. The researchers utilized six moving light bar patterns continuously and simultaneously between the laminar flows to achieve cell separation based on the size and electrical properties. The separated cell populations were then transferred and collected through the flow in two branches (Figure 9.9b).

The results demonstrated that this circulating tumor cell (CTC) separation scheme effectively isolated PC-3 cells or OEC-M1 cells from white blood cells, with high recovery rates (76–83% for OEC-M1 cells, 61–68% for PC-3 cells) and high purity (74–82% for PC-3 cells, 64–66% for OEC-M1 cells). Additionally, the viability of the separated cancer cells (PC-3 and OEC-M1) was maintained at high levels (94.2% and 95.3%, respectively). This label-free, continuous, and cell-friendly

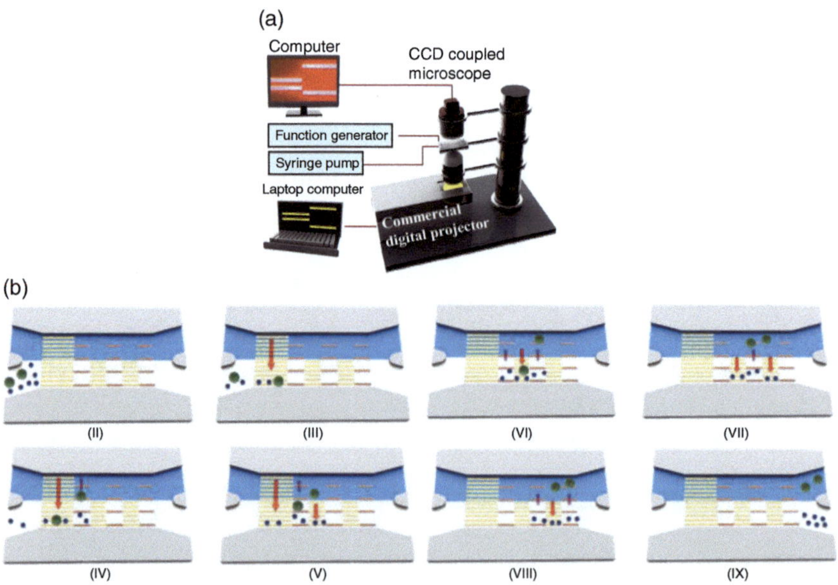

Figure 9.9 Schematic illustration of (a) the overall experimental setup and (b) the whole cancer cell isolation processes. *Source*: [18] / The Royal Society of Chemistry.

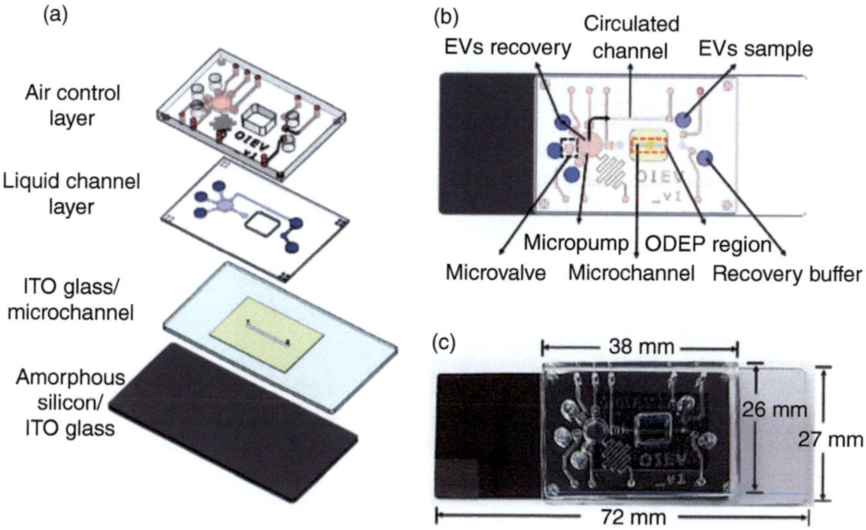

Figure 9.10 Design and microfabrication of the OIEV platform. (a) An exploded view of the optically-induced EV isolation (OIEV) platform; (b) top view of the platform; (c) a photograph of the OIEV platform and the microfluidic flow control part. *Source*: [19] / The Royal Society of Chemistry.

microfluidic separation method ensures the purity and viability of the separated cells, making it particularly valuable for subsequent cell-based or biochemical analyses. The ability to separate CTCs in a highly pure and viable manner holds a significant promise for advancing cancer research and diagnostics.

Chen et al. utilized polydimethylsiloxane (PDMS) to fabricate a pneumatic control layer above a sandwich structure, integrating various functional pneumatically actuated microfluidic components in their study [19]. These components included a micro-pump (flow actuator), multiple micro-valves (flow switches), a circulation channel (flow loop), and several microchambers for liquid loading and extracellular vesicle recovery (Figure 9.10a–c). The microfluidic system developed by the researchers enables label-free and contactless isolation and enrichment of extracellular vesicles on virtual electrodes.

The adjustable and dynamically varying virtual electrodes within the system allow for the rapid enrichment of extracellular vesicles, increasing their concentration by 272 times within a short period of 32 seconds. The pneumatic control device cyclically captures extracellular vesicles, achieving a high absorption rate with a release efficiency of 99.8% and a recovery rate of 52.2%. This innovative microfluidic system offers a highly efficient method for isolating and enriching extracellular vesicles, showcasing the potential of pneumatic control technology in microfluidics for biomedical applications.

Valley et al. introduced a novel approach to dynamically generate virtual electrodes within an OET system by employing patterned light illumination in their study [20]. These virtual electrodes serve a dual purpose: they enable the manipulation of multiple cells and facilitate the selective electroporation of individual cells over a wide area (Figure 9.11). The electroporation process was monitored using

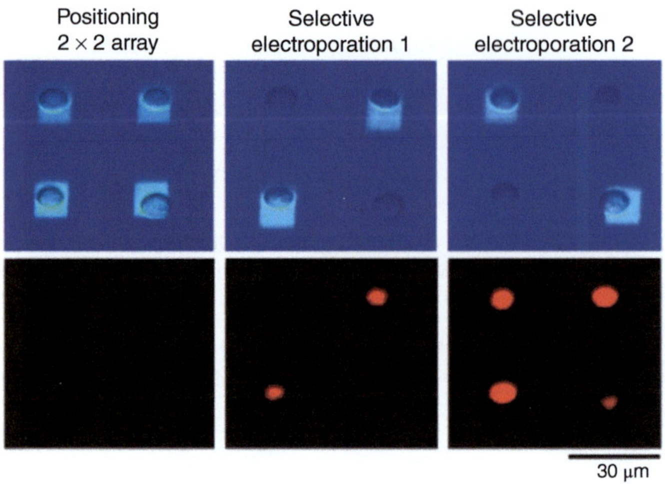

Figure 9.11 Parallel single-cell electroporation. *Source*: [20] / The Royal Society of Chemistry.

fluorescent dyes, allowing the researchers to observe how the DEP forces change in response to electroporation. Furthermore, the device integrated fluidic channels that enabled the exchange and perfusion of various media. These channels were instrumental in demonstrating the reversible electroporation of cells. Through the single-cell selective determination of optimal electroporation conditions, the researchers identified that the ideal electric field strength for electroporation, while maintaining cell viability, falls within the range of 1.4–2.3 kV/cm.

As mentioned earlier, exceeding a certain threshold of electric field intensity, such as 2.3 kV/cm, can result in the failure of pores induced in the cell membrane by electroporation to reseal autonomously, leading to cellular lysis. In response to this phenomenon, the OET technology has been developed for the photoelectric control of cellular lysis [21–25]. Lin et al. introduced an optically induced cell lysis device capable of selectively lysing individual cells within a cell population, a feat not achievable with conventional tools. This chip-level device, constructed from optoelectric materials, can generate nonuniform electric fields at specific locations when exposed to a light beam spot, leading to the creation of transmembrane potentials within cells. Through this method, cell lysis can be achieved via optically induced electric fields [21].

Figure 9.12a–d depicts the transition of fibroblasts at the single-cell level before and after the application of a nonuniform electric field [21]. Initially, the cells exhibited green fluorescence, indicating active biological vitality. However, upon exposure to the light beam, the cell membrane ruptured due to the photogenerated nonuniform electric field, resulting in the release of intracellular organelles and eventual cellular debris, confirming cell destruction. Despite inducing physical damage to the cell membrane, the nucleus remains unharmed [21].

Additionally, OET technology has shown the ability to selectively lyse various types of adherent or non-adherent cells in complex biological samples, including crowded environments, thereby overcoming a key challenge in the analysis of rare

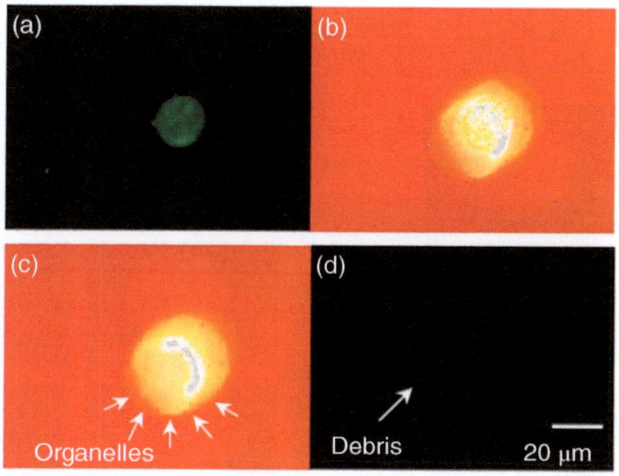

Figure 9.12 Sequence of images depicting the lysis process of an individual fibroblast cell. (a) Cell stained green; (b) light induces transmembrane potential; (c) cell disrupted, releasing organelles (black dots); (d) cell fully disrupted after light is blocked. *Source*: [21] / AIP Publishing LLC.

cells or specific cell populations. In the study of Witte et al. [22], an amorphous silicon (a-Si) photoconductor electrode is integrated into the microfluidic environment, enabling the selection and lysis of both suspension and adherent cells under physiological conditions – an advancement over previous studies. The integration of microchannels and the reduction of beam sizes not only enhance spatial selection but also significantly lower the power densities required, five orders of magnitude less than previous methods, ensuring minimal damage to cells. The study demonstrates that light-patterned electrical fields can effectively lyse single cells within dense populations and under physiological conditions. The precise control achieved using the small optical spot and integrated photoconductor allows for selective lysis within one minute, applicable to both non-adherent cells like red blood cells (RBCs) and adherent cells like michigan cancer foundation-7 (MCF7) cells.

In addition to traditional OET devices, a microfluidic device integrating OET and cell counting functions has been developed for cellular lysis [23]. This device utilizes hydrodynamic forces to concentrate cells at the center of the microfluidic channel, where cell counting is performed using integrated optical fibers. Subsequently, an electric field is induced by projected optical images to achieve continuous cellular lysis. Experimental results have demonstrated the device's accuracy in cell counting and its ability to successfully lyse 93.8% of the cells. This integrated microfluidic system represents a promising approach for efficient and selective cellular lysis in biological and biomedical applications.

Huang et al. [24] introduced an approach utilizing OET to enhance the extraction and collection of cell nuclei through optically induced cell lysis (OICL) in an automated format (Figure 9.13). Compared to existing nucleus extraction methods, this innovative OICL device offers superior flexibility, selectivity, and efficiency. By integrating optically-induced dielectrophoretic (ODEP) with the OICL module within a single microfluidic chip, the need for highly trained personnel and expensive equipment is eliminated.

This microfluidic device innovatively combines OICL and ODEP for selective cell membrane lysis and nucleus separation, marking the first demonstration of such integration for continuous-flow nucleus extraction. The device facilitates the lysis of

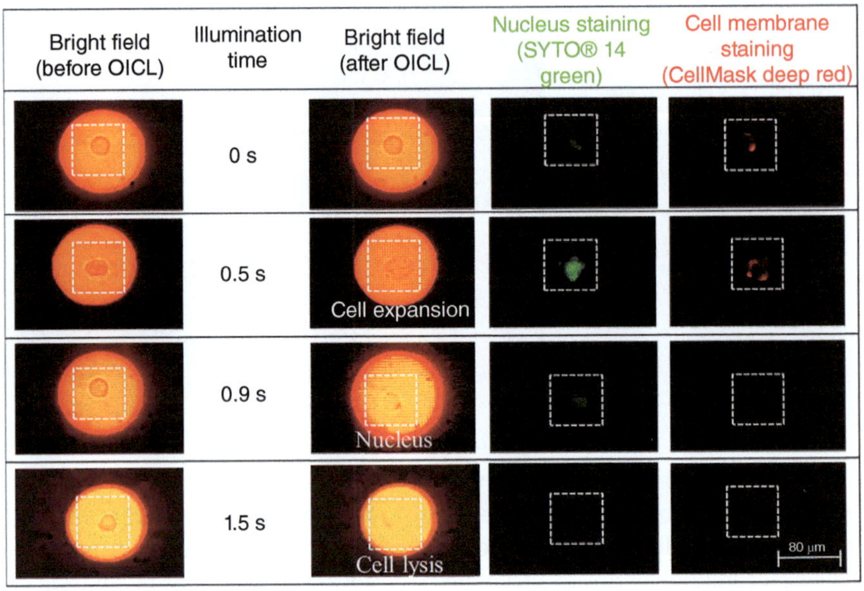

Figure 9.13 The effect of illumination time on cell membrane lysis and expansion. *Source*: [24] / The Royal Society of Chemistry.

cell membranes without disrupting nuclei and enables the separation of intact nuclei from cellular debris. It shows promise for applications in nuclear transfer, genetic replication, and animal cloning studies.

References

1 Yu, P., Ran, J., Yang, R. et al. (2024). Rapid isolation of pan-neutralizing antibodies against Omicron variants from convalescent individuals infected with SARS-CoV-2. *Frontiers in Immunology* 15: 1374913.

2 Le, K., Tan, C., Le, H. et al. (2020). Assuring clonality on the beacon digital cell line development platform. *Biotechnology Journal* 15 (1): e1900247.

3 Wang, L.T., Cooper, A.J., Farrell, B. et al. (2024). Natural malaria infection elicits rare but potent neutralizing antibodies to the blood-stage antigen RH5. *Cell* 187 (18): 4981–4995 e14.

4 Zhang, Y., Park, M., Ghoda, L.Y. et al. (2024). IL1RAP-specific T cell engager depletes acute myeloid leukemia stem cells. *Journal of Hematology & Oncology* 17 (1): 67.

5 Wei, J., Han, X., Bo, J., and Han, W. (2019). Target selection for CAR-T therapy. *Journal of Hematology & Oncology* 12 (1): 62.

6 Liu, Y., Yan, X., Zhang, F. et al. (2021). TCR-T Immunotherapy: the challenges and solutions. *Frontiers in Oncology* 11: 794183.

7 Nowicki, T.S., Peters, C.W., Quiros, C. et al. (2023). Infusion product TNFalpha, Th2, and STAT3 activities are associated with clinical responses to transgenic T-cell receptor cell therapy. *Cancer Immunology Research* 11 (12): 1589–1597.

8 Cruz-Ramos, M. and Garcia-Foncillas, J. (2019). CAR-T cell and personalized medicine. *Advances in Experimental Medicine and Biology* 1168: 131–145.

9 Shafer, P., Kelly, L.M., and Hoyos, V. (2022). Cancer therapy with TCR-engineered T cells: current strategies, challenges, and prospects. *Frontiers in Immunology* 13: 835762.

10 Dolton, G., Tungatt, K., Lloyd, A. et al. (2015). More tricks with tetramers: a practical guide to staining T cells with peptide-MHC multimers. *Immunology* 146 (1): 11–22.

11 Bunse, L., Green, E.W., and Platten, M. (2019). High-throughput discovery of cancer-targeting TCRs. *Methods in Enzymology* 629: 401–417.

12 Ran, T., Eichmüller, S.B., Schmidt, P., and Schlander, M. (2020). Cost of decentralized CAR T-cell production in an academic nonprofit setting. *International Journal of Cancer* 147 (12): 3438–3445.

13 Gan, C., Liang, S., Wang, F. et al. Non-contact massively parallel manipulation of micro-objects by optoelectronic tweezers. *2021 WRC Symposium on Advanced Robotics and Automation (WRC SARA)* (19 November 2021), IEEE, pp. 7–12.

14 Gan, C., Zhang, J., Zhao, J. et al. (2022). Cell focusing and rotation by a planar optoelectronic tweezers. *2022 International Conference on Manipulation, Automation and Robotics at Small Scales (MARSS).* (2 September), IEEE, pp. 1–5.

15 Zheng, Y., Nguyen, J., Wei, Y., and Sun, Y. (2013). Recent advances in microfluidic techniques for single-cell biophysical characterization. *Lab on a Chip* 13 (13): 2464–2483.

16 Liang, S., Gan, C., Dai, Y. et al. (2021). Interaction between positive and negative dielectric microparticles/microorganism in optoelectronic tweezers. *Lab on a Chip* 21 (22): 4379–4389.

17 Liang, S., Sun, J., Zhang, C. et al. (2022). Parallel manipulation and flexible assembly of micro-spiral via optoelectronic tweezers. *Frontiers in Bioengineering and Biotechnology* 10: 868821.

18 Huang, S.B., Wu, M.H., Lin, Y.H. et al. (2013). High-purity and label-free isolation of circulating tumor cells (CTCs) in a microfluidic platform by using optically-induced-dielectrophorctic (ODEP) force. *Lab on a Chip* 13 (7): 1371–1383.

19 Chen, Y.S., Lai, C.P.K., Chen, C., and Lee, G.B. (2021). Isolation and recovery of extracellular vesicles using optically-induced dielectrophoresis on an integrated microfluidic platform. *Lab on a Chip* 21 (8): 1475–1483.

20 Valley, J.K., Neale, S., Hsu, H.Y. et al. (2009). Parallel single-cell light-induced electroporation and dielectrophoretic manipulation. *Lab on a Chip* 9 (12): 1714–1720.

21 Lin, Y.-H. and Lee, G.-B. (2009). An optically induced cell lysis device using dielectrophoresis. *Applied Physics Letters* 94 (3).

22 Witte, C., Kremer, C., Chanasakulniyom, M. et al. (2014). Spatially selecting a single cell for lysis using light-induced electric fields. *Small* 10 (15): 3026–3031.

23 Lin, Y.-H. and Lee, G.-B. (2010). An integrated cell counting and continuous cell lysis device using an optically induced electric field. *Sensors and Actuators B: Chemical* 145 (2): 854–860.

24 Huang, S.H., Hung, L.Y., and Lee, G.B. (2016). Continuous nucleus extraction by optically-induced cell lysis on a batch-type microfluidic platform. *Lab on a Chip* 16 (8): 1447–1456.

25 Kremer, C., Witte, C., Neale, C.L. et al. (2014). Shape-dependent optoelectronic cell lysis. *Angewandte Chemie (International Ed. in English)* 53 (3): 842–846.

10

Challenges and Future Directions

10.1 Current Challenges

While OETs have been demonstrated in several valuable applications, significant hurdles remain in transitioning this technology from a predominantly laboratory-based technique to one that is widely adopted for practical use. Therefore, further developments and enhancements are crucial to overcoming current limitations in the state of the art.

The performance of OET devices is heavily dependent on the photoconductive materials used. At present, a-Si:H is favored for constructing OET's photoconductive layer due to its high stability, effective light absorption, durability, and biocompatibility. However, producing a-Si:H is a complex and costly endeavor, necessitating the skilled operation of plasma-enhanced chemical vapor deposition (PECVD) within a cleanroom environment, using optimized parameters such as temperature and pressure. On the other hand, alternative organic photoconductive materials like poly(3-hexylthiophene-2,5-diyl)/[6,6]-phenyl-C61-butyric acid methyl ester (P3HT/PCBM) polymers and TiOPc offer simpler fabrication through spin-coating, but they suffer from reduced lifespan and stability when exposed to moisture, thus presenting additional challenges.

A significant limiter for a-Si:H-based OET systems is the necessity for low-conductivity liquids to function effectively. This poses a major challenge for biomedical applications, particularly where high-conductivity, cell-friendly buffers are involved. Patterned phototransistors can address this issue; however, their fabrication remains a formidable challenge, involving sophisticated techniques such as ion implantation, photolithography, and multi-step plasma etching. Presently, only a handful of research groups globally possess the expertise and facilities to produce phototransistor-based OET devices.

Furthermore, although companies like Berkeley Lights have made strides by commercializing automated OET systems such as the Beacon and Lightning systems, the associated costs make the widespread deployment of phototransistor-based OET devices impractical for certain uses. The company mitigates contamination risks by designing single-use chips for these systems. Nonetheless, the high expense highlights the urgent need for new photoconductive materials and innovative

Optoelectronic Tweezers for Manipulation at the Micro and Nano Scale, First Edition.
Lin Feng, Menglu Tan, Ao Wang, and Jiaying Zhang.
© 2026 WILEY-VCH GmbH. All rights reserved, including rights for text and data mining and training of artificial intelligence technologies or similar technologies. Published 2026 by WILEY-VCH GmbH.

phototransistor structures that can overcome these challenges and expand OET's applicability further.

Additionally, OET is largely characterized as a two-dimensional manipulation technique, which presents a downside compared to optical tweezer (OT), especially since reliable three-dimensional manipulation with OET is still not feasible. Addressing this shortcoming will likely necessitate innovating new devices or related technologies that support 3D manipulation. Moreover, the majority of OET research, approximately 90%, has concentrated on vertical OET setups, which imposes constraints on the technology's versatility and range of applications. Exploring three-dimensional manipulation and diversifying device formats could significantly broaden OET's scope and utility in future research and applications.

10.2 Innovation and Ongoing Research

The future development of OET technology holds immense promise for driving transformative advancements across a wide spectrum of scientific disciplines. At the forefront of this evolution, researchers are actively pursuing strategies to enhance the integration and intelligence of OET systems, aiming to create more adaptable and versatile platforms capable of handling increasingly complex tasks [1–6]. This includes the development of advanced control algorithms and system architectures that can dynamically respond to varying experimental conditions, thereby expanding the operational flexibility of OET systems.

One of the most significant areas of progress lies in overcoming the current limitations of OET technology to enable in vivo applications. This breakthrough would dramatically expand the potential impact of OET in biological and medical research, opening new avenues for minimally invasive cellular manipulation and targeted drug delivery within living organisms. Researchers are making substantial strides in this direction by developing biocompatible materials and optimizing optical parameters to ensure safe and effective operation in physiological environments.

The exploration of novel photoconductive materials represents another crucial frontier in OET technology development. Recent advancements in material science have led to the discovery and synthesis of materials with enhanced photoconductive properties, improved durability, and greater sensitivity to specific wavelengths of light. These innovations are expected to significantly boost the performance and reliability of OET systems, enabling more precise and efficient manipulation of microscopic particles and biological cells.

Furthermore, the integration of automation and machine learning technologies is poised to revolutionize OET capabilities. Advanced imaging recognition algorithms combined with intelligent path planning systems are being developed to enable autonomous and precise manipulation on demand. This integration allows for real-time analysis and adaptive control of OET operations, significantly enhancing the technology's efficiency and accuracy in complex experimental scenarios.

The continuous improvement and progression of OET technology are anticipated to stimulate the development of innovative OET-driven micromachines.

These sophisticated devices, capable of performing intricate tasks at the microscale, are expected to pave the way for novel applications in fields ranging from targeted therapeutics to micro-robotics. However, achieving these advancements requires sustained interdisciplinary collaboration and united efforts among researchers worldwide, combining expertise from fields such as optics, materials science, biology, and computer engineering.

It is particularly noteworthy to mention that the development of OT and OET technologies has catalyzed the emergence and propelled the advancement of optically driven machinery. This symbiotic relationship has created a mutually reinforcing dynamic between optical manipulation techniques and optically driven machinery. As progress in optically driven machinery continues, it reciprocally facilitates advancements in optical manipulation techniques, unlocking their full potential and giving rise to numerous new applications in biology, physics, and biomedical engineering.

10.3 Future Development Direction and Suggestions

To further advance OET technology, rigorous investigation into the foundational physics that governs its operations is essential. Understanding the intricate interactions and behaviors of targets under different experimental conditions remains a significant challenge. Central to the OET functionality are several key mechanisms activated by light, including dielectrophoresis (DEP), alternating current electroosmosis (ACEO), AC electrothermal (ACET) effects, and electrodeposition. While these mechanisms have been individually studied and characterized, they do not operate in isolation in practical scenarios. Instead, they interact in complex, often unpredictable ways that existing models struggle to encapsulate. Hence, developing a comprehensive theoretical model that integrates all these mechanisms is crucial. Such a model would enable the accurate prediction and explanation of target behaviors during OET manipulation. Furthermore, unexplained phenomena, such as the spontaneous rotation of certain cells, the three-dimensional movement of micro-objects, and electrohydrodynamic interactions between polarized micro-objects, highlight the need for deeper research. Understanding these phenomena is vital to unlocking the full potential of OET in diverse applications.

In addition to refining the theoretical framework, expanding the scope of the OET hardware is imperative. Significant potential exists in lateral OET devices, which have already shown improved capabilities in assembling micro- and nano-optoelectronic elements, such as micro-disk lasers and nanowires. Moreover, the field of optoelectrowetting (OEW) remains underexplored, offering a fertile ground for research and potential breakthroughs. Currently, OET applications are confined to small areas, typically less than $2\,cm^2$, limiting their use in broader industrial contexts. Efforts should focus on expanding these applications to larger areas to facilitate manufacturing processes. One of the persistent challenges in OET technology is the nonspecific adhesion of cells and microparticles to the device surfaces, which hampers performance and accuracy. To address this, researchers must develop specialized coatings that prevent unwanted binding without compromising the device's effectiveness.

The integration of OET with other advanced analytical and measurement tools is another critical area for growth. By combining OET systems with established techniques like atomic force microscopy, confocal microscopy, and Raman spectroscopy, researchers can achieve a more holistic view of micro- and nanoscale phenomena. This integration would allow for the simultaneous acquisition and analysis of multiple parameters, thus providing deeper insights and more comprehensive data.

Moreover, the prohibitive cost of current automated OET systems, such as those offered by Berkeley Lights, presents a barrier to widespread adoption, particularly in academia and smaller research labs. While these systems, like the Beacon, are primarily used in large pharmaceutical companies due to their expense, there is a clear need for more cost-effective solutions. The introduction of the Lightning system, with reduced throughput and automation but at a lower price point, is a positive development. However, further efforts are needed to create affordable, fully automated OET systems that can facilitate research across various disciplines. Complementing these hardware advancements, the development of sophisticated software and algorithms is necessary to enhance the functionalities of OET systems. These advancements could include automated image acquisition, target identification, and precise manipulation, ultimately resulting in a highly efficient and versatile tool.

In summary, by addressing these challenges and pursuing these avenues for development, OET research can be richly expanded. As the technology progresses, it is poised to play a crucial role in materials science, biology, and pharmaceuticals, offering transformative opportunities for research and innovation in these fields. With a comprehensive approach that includes both theoretical and practical advances, OET systems can become even more powerful and widely used tools in scientific exploration and industrial applications.

References

1 Wang, A., Gan, C., Han, H. et al. 2024. Dynamic adaptive imaging system on optoelectronic tweezers platform. *IEEE International Conference on Robotics and Automation (ICRA)* (13–17 May), IEEE, pp. 15622–15627.

2 Zhao, J., Gan, C., Zhang, J. et al. 2022. Deep learning assisted automated separation platform of single cells and microparticles using optoelectronic tweezers. *2022 WRC Symposium on Advanced Robotics and Automation (WRC SARA)* (20–20 August), IEEE, pp. 275–279.

3 Liu, J., Wang, H., Liu, M. et al. (2022). POMDP-based real-time path planning for manipulation of multiple microparticles via optoelectronic tweezers. *Cyborg and Bionic Systems* 2022: 9890607.

4 Siu, D.M., Lee, K.C., Chung, B.M. et al. (2023). Optofluidic imaging meets deep learning: from merging to emerging. *Lab on a Chip* 23 (5): 1011–1033.

5 Hayashi, M., Ohnuki, S., Tsai, Y. et al. (2023). Is AI essential? Examining the need for deep learning in image-activated sorting of Saccharomyces cerevisiae. *Lab on a Chip* 23 (19): 4232–4244.
6 Huang, K., Matsumura, H., Zhao, Y. et al. (2022). Deep imaging flow cytometry. *Lab on a Chip* 22 (5): 876–889.

Index

Optoelectronic Tweezers for Manipulation at the Micro and Nano Scale, First Edition.
Lin Feng, Menglu Tan, Ao Wang, and Jiaying Zhang.
© 2026 WILEY-VCH GmbH. All rights reserved, including rights for text and data mining and training of artificial intelligence technologies or similar technologies. Published 2026 by WILEY-VCH GmbH.